Atom Vibrations in Solids: Amplitudes and Frequencies

I0028385

CAMBRIDGE SCIENTIFIC PUBLISHERS

ATOM VIBRATIONS IN SOLIDS: AMPLITUDES AND FREQUENCIES

Contents

i

Physics Reviews 2004, Vol 21, pp. 1-162
Reprints available directly from publisher
Printed in UK
Photocopying permitted by license only

ATOM VIBRATIONS IN SOLIDS: AMPLITUDES AND FREQUENCIES

V.V. Levitin

Zaporozhye State Technical University,
Zaporozhye 69063, Ukraine

ABSTRACT

Many properties of solids are related to crystal lattice oscillations. This phenomenon is of great theoretical and practical significance. This review is devoted to the data of experimental investigations of heat atom vibrations: it contains specific physical values, which have been measured directly and also includes the methods for X-ray measurements of the mean-square atom displacements. Techniques of the frequency spectra determination with neutrons scattering are described together with a review and analysis of quantitative data on amplitudes of the atom vibrations for metals, solid solutions, semiconductors, intermetallic and chemical compounds, carbides. The effect of composition and temperature upon vibrations is interpreted. Frequency spectra of real solids are presented and interplanar atomic force constants are described. The review discusses velocities of crystal lattice waves and considers the anharmonity of oscillations.

PREFACE

Problems of atomic micro-structure are basic in solid state physics. The atoms and ions which are bonded with each other with considerable interatomic forces are not motionless. Due to the consistent vibrating movements, they are permanently deviating from their equilibrium position. Elastic waves of different lengths, frequencies, and amplitudes run through crystalline solids at all times. The typical order of the atomic vibrations frequencies is 10^{13} Hz, and that of the amplitudes is 10^{-11} m.

The phenomena of atomic vibrations reflecting the interaction of micro-particles with each other depend on the deep properties of the medium; hence, they are of great theoretical interest. Investigation of the atoms motion provides the researcher with an efficient tool to study the specific features of the bodies structure. The problem of wave atomic displacements is also important from the practical viewpoint. The amplitude-frequency characteristics of the vibrating spectrum of the alloy can be varied, for example by alloying, to produce a well-directed effect on the properties of the materials. The specific features of the vibrating motion determine a number of properties: strength at high temperatures, conductivity, heat capacity, thermal expansion, sound speed, melting point.

The process of the atomic vibrations is important for materials of different classes: for metallic, covalent, ionic crystals, semiconductors, intermetallic compounds, interstitial phases.

The problem of the vibrating atomic motion in solids is significant and has a number of experimental, theoretical and applied aspects. Naturally, it is necessary to identify the scope of problems considered in this book.

The author presents mainly experimental data obtained with the help of modern methods of research, and attention is paid to the

3

parameters of the dynamics of the crystal lattice which are directly measured in the experiments.

The book will be useful for researchers, engineers, university students, post-graduate students, companies workers. The book is intended for all those who are interested in the problems of interatomic interactions, in scientific foundations of constructing materials with prescribed properties, in material studying aspects of the industry, in physical nature of the strength, heat, electrical and other properties of solids.

It is known that the human
knowledge deserves the name of
Science depending on the role
played there by the number.

Émile Borel, French mathematician

INTRODUCTION

The first studies in the area of dynamics of crystal lattice were completed at the beginning of the 20th century by A. Einstein, M. Born, T. Karman, P. Debye, I. Waller, L. Brillouin and their collaborators. They studied the basic laws of the elastic wave propagation in the periodic structures, described the quantum mechanical peculiarities of the wave motion and investigated the influence of the atomic displacements on scattering of X-rays by solids.

The classical theories gave impetus to further investigations of elastic waves in solids. Beginning in the 1950s, scientific studies progressed in several directions. These directions are interconnected by the common object of investigation but are different in the methodological approach and the goals formulated. The numerous investigations fulfilled and the papers published can be grouped in the following way.

The first group is devoted to the measurements of the atomic displacement amplitudes on the basis of determination of the X-rays scattering intensity. Such measurements have been made for different temperatures both below and above room temperature. The obtained mean-square values of the amplitudes were treated, in particular, as a measure of the interatomic bonding forces.

The studies, in which the problem of determination of the frequency dependence of the elastic vibration modes on the wave vector (or, what is the same, on the elastic wave length in the crystal lattice) is considered, can be included in the second group of investigations. This problem has been solved by using the scattering of thermal neutrons by monocrystals. On the base of the experimental data, one can find the frequency spectrum of real solids.

The third group of investigations is the study of the parameters of anharmonic atomic vibrations, i.e., the characteristics of the deviation of the oscillations from the harmonic law.

The fourth group can be composed of the calculations of the parameters of the lattice dynamics with the help of computers. The calculations of the frequency spectra, dispersion curves, and simulations can be included into this group.

The theory of vibrating processes and wave propagation in an ideal crystal lattice became classical and is considered in a number of monographs, surveys, and text-books. In particular, a detailed presentation of these problems can be found in [1]–[7].

Among the books specially devoted to lattice dynamics, one can distinguish "Theory of Lattice Dynamics in the Harmonic Approximation" by A.A. Maradudin, E.W. Montroll, *et al.* [1] and the monograph by J. Reisland "Physics of Phonons" [2]. These books are written by theoreticians and are intended mainly for the theoreticians. The authors give experimental data mainly to illustrate the theoretical statements. The extremely high scientific level and, presumably, timeless value of the book by J. Reisland should be noted.

This review is designed mainly to cover the systematic experimental data accumulated in the last decades when studying the crystal lattice dynamics for various materials.

The logic of presentation require the consideration of a number of problems closely connected with the main ones. At the beginning of the book, in the first Section, a range of basic concepts and representations of crystal lattice dynamics is introduced. Here, the harmonic oscillations of one-dimensional atomic chains are considered as model objects; the processes of the elastic wave propagation in three-dimensional crystals are described. The main goal of this section is to prepare the reader for the study of the next sections.

The second section is devoted to methodological problems. Here, methods for experimental determination of the mean-square amplitudes of the atomic displacements and approaches to constructing the dependence of the oscillatory modes densities of the crystal on the frequency are described.

In the next section, the results of determination of the atomic vibration amplitudes in metals are presented; the influence of the alloy composition and of the temperature as well as the phenomenon of the vibration anisotropy are considered.

The fourth section contains the data of atomic amplitudes measurements in semiconductor materials and in chemical compounds. Here, the connection between the amplitudes of the atomic vibrations and the properties of the phases under consideration is analyzed.

The description and the analysis of the frequency spectra for various solids is the subject of the fifth section.

In the final section, problems of anharmonicity of vibrations of real crystal bodies are presented. The connection between the anharmonicity parameters and the structure of the solid is discussed.

The objective of the review is to analyze the experimentally determined data and so the Debye model, which played an important role in the development of the theory of crystal lattice dynamics, is scarcely used. For this reason, we try to avoid data concerning the determination of the Debye temperature of elements and compositions, preferring to use the mean-square amplitude of the atomic vibrations measured in the experiments. When considering the vibration frequency spectrum, the real determined spectrum was also preferable over the spectrum of the Debye model.

1. HARMONIC VIBRATIONS OF THE CRYSTAL LATTICE OF SOLIDS

1.1. Main concepts of the wave motion

The wave motion can be characterized by several important parameters: amplitude, wave length, frequency. The equation of the wave propagation in a fixed medium represents a dependence between the deviation u of a chosen point relative its equilibrium position and its coordinate x and the time t. The general expression of this dependence for a plane harmonic wave has the following form:

$$u = A \exp \left[i \left(\omega t - \frac{2\pi}{\lambda} x \right) \right], \tag{1.1}$$

where A is the amplitude, m; u is the displacement from the equilibrium position, m; ω is the angular frequency, rad/s; λ is the wave length, m. The angular frequency is connected with the frequency ν, i.e., the number of oscillations per time unit, by the relation $\omega = 2\pi\nu$.

The quantity $q = 2\pi/\lambda$ is called the module of the wave vector; this vector is directed along the wave propagation.

It is known that the function $\exp(i\varphi)$ is periodic with the period 2π: $\exp(i\varphi) = \exp[i(\varphi + 2\pi n)]$, where n is an integer. Therefore, we see that the wave displacement of a point is doubly periodic: with respect to time and coordinate, where the time period is $T = 2\pi/\omega$ and the coordinate one is λ. Expression (1.1) is complex. Generally speaking, one should write the symbol $\Re$ before the right-hand side, since only the real part of the formula has a physical sense. One can also use for the wave equation the function sine or cosine of the argument written in the parenthesis. However, the relation in the exponential form (1.1) is more convenient, especially, when summing waves with different amplitudes and phases. The passage from one form of the equation to the other one can be fulfilled with the help

9

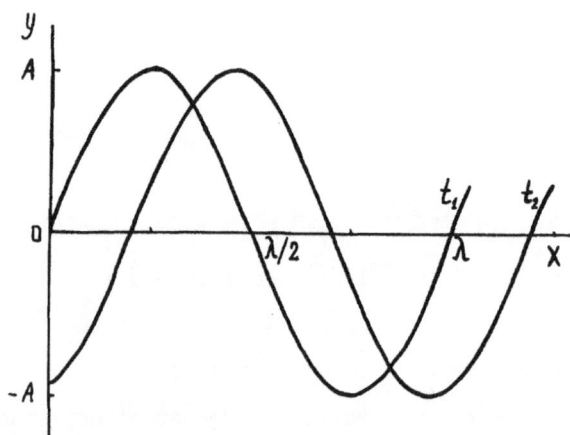

FIGURE 1.1. The graph of the traveling wave: the two curves correspond to two successive time intervals; $t_2 > t_1$.

of the Euler formula

$$\exp(i\varphi) = \cos\varphi + i\sin\varphi. \tag{1.2}$$

The graph of the transverse traveling wave is shown in Fig. 1.1

When the traveling wave reflects from the interface of two media, a standing wave is formed as a result of interaction of the direct and inverse waves. The specific features of the standing wave are the facts that nodes and antinodes appear and the one-way transfer of the energy in the medium is absent. The equation of the standing wave can be obtained by summing the real parts of two equations (1.1) with the opposite signs of the second term in the exponent:

$$u = 2A\left(\sin\omega t\right)\left(\cos\frac{2\pi}{\lambda}x\right). \tag{1.3}$$

The graph of the standing wave (Fig. 1.2) shows the evolution of the displacements in time.

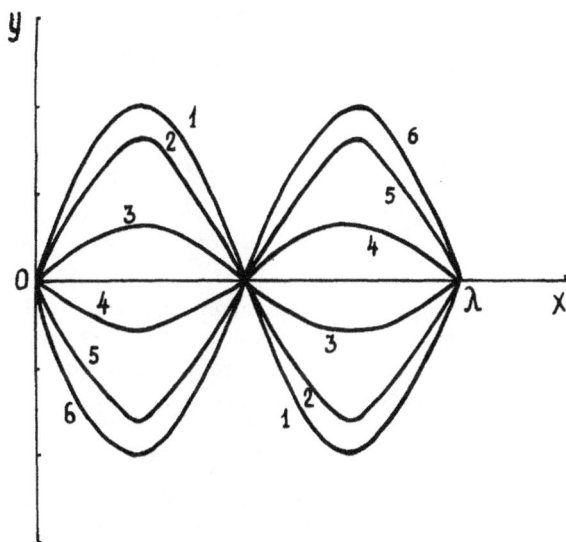

FIGURE 1.2. The graph of the standing wave: 1–6 are the positions of the displacements in successive time intervals: 0; 0.1 T; 0.2 T; ... ; 0.5 T.

1.2. Vibrations of one-dimensional chain with atoms of equal mass

It is useful to consider the complicated and various processes of collective atomic vibrations with simple models. It occurs that the one-dimensional models of atomic sequences have many characteristic features appropriate for a real three-dimensional crystal. On one hand, the relative simplicity of phenomena and, on the other hand, the essential character of conclusions justify the attention paid to the wave propagation in a linear crystal.

Let an infinitely long straight-line chain consisting of identical atoms of mass m be given, in which the atoms are equally spaced at intervals a (Fig. 1.3a). We take an arbitrary particle as the zero one, the next is the first one, the next is the second and so on. The atoms are bonded with each other by forces of interatomic interaction. In the equilibrium state, the kinetic energy of the atomic chain is equal to zero, and its potential energy has a minimum; accordingly, the

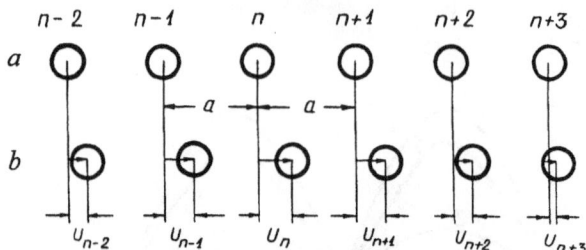

FIGURE 1.3. Vibrations of the infinite one-dimensional atomic chain: a) — the atoms are in the equilibrium position; b) — the atoms are displaced from the equilibrium position due to passing a longitudinal wave; the vectors of the displacements u_{n-2}, u_{n-1}, u_n, ... are shown.

total force acting on each atom is equal to zero. Thus, the numbers of the atoms are 0; 1; 2; 3; ...; n; $n+1$; $n+2$; $n+3$; .... .

Strictly speaking, when considering an infinitely long chain, some physical parameters become indefinite (for instance, the total energy of vibrations and also the length). For a chain of a finite length, the boundary conditions should be taken into account by introducing additional forces acting on the boundary atoms in order to preserve the structure of the one-dimensional lattice. However, it is easier to assume that the length of the chain is essentially greater than the interatomic distance and the forces of interatomic interaction decrease rather quickly in the distance.

Let us consider the case of longitudinal vibrations. Let the wave perturbation deviate the atoms from the equilibrium position (Fig. 1.3b). The atoms are deviated by the length of intervals denoted by u: the nth atom is displaced an amount u_n; the $n+1$th, u_{n+1} and so on.

One can easily obtain an expression for the kinetic energy of a vibrating crystal. For this purpose, one should sum up the kinetic energies of all atoms:

$$T = \frac{1}{2} \sum_{n=0}^{N} m\dot{u}_n^2, \qquad (1.4)$$

where $\dot{u}_n$ is the velocity of an nth atom. The summation is carried over all atoms of the model.

Contrary to the kinetic energy, the potential one is a function of the mutual location of the atoms. The total potential energy depends only on the distance between the interacting atoms. For instance, consider two atoms: the atom number i and the atom number $i + p$, where p is a positive integer. The coordinate of the second atom is $x = pa$. The potential energy of interaction of these two atoms can be written as

$$\Phi(r_{i,i+p}) = \Phi(x_{i+p} - x_i), \qquad (1.5)$$

where the parenthesis denotes a functional dependence. To obtain the value of the total potential energy of an one-dimensional chain, one has to sum up the interaction energy for all pairs of atoms:

$$\Phi = \sum_{i=0}^{n-1} \sum_{p=1}^{n-i} \Phi(x_{i+p} - x_i). \qquad (1.6)$$

The number p in the internal sum varies from 1 to $n - i$ and the value i in the external sum varies from 0 to $n - 1$. This enables us to account once the interaction of all atomic pairs. For instance, the potential energy of a chain consisting of six atoms has the following form:

$$\begin{aligned}
\Phi = &\left[\Phi(x_1 - x_0) + \Phi(x_2 - x_0) + \Phi(x_3 - x_0) + \Phi(x_4 - x_0)\right.\\
&\left. + \Phi(x_5 - x_0)\right] + \left[\Phi(x_2 - x_1) + \Phi(x_3 - x_1) + \Phi(x_4 - x_1)\right.\\
&\left. + \Phi(x_5 - x_1)\right] + \left[\Phi(x_3 - x_2) + \Phi(x_4 - x_2) + \Phi(x_5 - x_2)\right]\\
&+ \left[\Phi(x_4 - x_3) + \Phi(x_5 - x_3)\right] + \left[\Phi(x_5 - x_4)\right].
\end{aligned}$$

One can easily see that the double sum (1.6) contains $n(n+1)/2$ terms, because each of $n + 1$ atoms interacts with all the others.

Let the relative displacements $(u_{i+p} - u_i)$ be small compared with the interatomic distance. Expand the function Φ into the Taylor series near the point $x = pa$ in powers of atomic displacements, restricting ourselves to the three terms of the series:

$$\begin{aligned}
\Phi(x_{i+p} - x_i) = \Phi(pa) &+ \left(\frac{d\Phi}{du}\right)_{x=pa} (u_{i+p} - u_i)\\
&+ \frac{1}{2}\left(\frac{d^2\Phi}{du^2}\right)_{x=pa} (u_{i+p} - u_i)^2 + \dots. \quad (1.7)
\end{aligned}$$

The derivatives are calculated at the point $x = pa$. Denote these derivatives $\Phi'(pa)$, $\Phi''(pa)$, ... Substituting expressions (1.7) into

(1.6), we obtain, omitting the powers higher than square:

$$\Phi = \sum_{i=0}^{n-1} \sum_{p=1}^{n-i} \Big[\Phi(pa) + \Phi'(pa)(u_{i+p} - u_i)$$

$$+ \frac{1}{2} \Phi''(pa)(u_{i+p} - u_i)^2 \Big]. \quad (1.8)$$

Thus, we have an expression for the total potential energy of the one-dimensional crystal lattice. The force, acting on one atom, say number r, is equal to minus gradient of the potential energy:

$$F_r = -\frac{d\Phi}{du_r}. \quad (1.9)$$

Hence, one has to differentiate the double sum (1.8) with respect to the displacement of the rth atom; only the terms with $i = r$ or $i + p = r$ make a contribution in the sum; all other terms do not depend on u_r. Let us see the table of displacements for our example of the chain of six atoms:

$u_1 - u_0$	$u_2 - u_0$	$\underline{u_3 - u_0}$	$u_4 - u_0$	$u_5 - u_0$
$u_2 - u_1$	$\underline{u_3 - u_1}$	$u_4 - u_1$	$u_5 - u_1$	
$\underline{u_3 - u_2}$	$u_4 - u_2$	$u_5 - u_2$		
$\underline{u_4 - u_3}$	$\underline{u_5 - u_3}$			
$u_5 - u_4$				

Consider, for example, the force acting on the third atom, $r = 3$. Then differentiating (1.8), we obtain non-zero result only for underlined terms; the terms with $i < r$ take into account the action of atoms positioned left from the chosen one; the terms with $i + p > r$ take into account the action of atoms positioned right from the chosen one. Thus, only the terms with $i = r$ or $i + p = r$ remain in the sums. The summation over i is not needed now. Regrouping the terms, we have (in formulae below, we assume that $n = 2r$; this assumption is justified by physical reasons)

$$
\begin{aligned}
F_r = & -\frac{\partial}{\partial u_r} \sum_{i=0}^{n-1} \sum_{p=1}^{n-i} \Big[\Phi(pa) + \Phi'(pa)(u_{i+p} - u_i) \\
& + \frac{1}{2}\Phi''(pa)(u_{i+p} - u_i)^2\Big] \\
= & -\frac{\partial}{\partial u_r} \sum_{p=1}^{n-r} \Big[\Phi'(pa)(u_{r+p} - u_r) + \frac{1}{2}\Phi''(pa)(u_{r+p} - u_r)^2 \\
& + \Phi'(pa)(u_r - u_{r-p}) + \frac{1}{2}\Phi''(pa)(u_r - u_{r-p})^2\Big] \\
= & -\sum_{p=1}^{n-r}\Big[-\Phi'(pa) - \Phi''(pa)(u_{r+p} - u_r) + \Phi'(pa) \\
& + \Phi''(pa)(u_r - u_{r-p})\Big]
\end{aligned}
\tag{1.10}
$$

Finally, we obtain the following expression for the force acting on the chosen atom by all other atoms as a result of their displacements from the equilibrium positions:

$$
F_r = \sum_{p=1}^{s} \Phi''(pa)\,(u_{r+p} + u_{r-p} - 2u_r). \tag{1.11}
$$

Here, $2s$ is the number of atoms interacting with the atom r under consideration.

The quantities $\Phi''(pa) = \left(\dfrac{\partial^2 \Phi}{\partial u^2}\right)_{x=pa}$ are called *atomic force constants*. Using expression (1.11), by the second Newton law, we obtain the equation of motion for the rth atom:

$$
m\frac{\partial^2 u_r}{\partial t^2} = \sum_{p=1}^{n-r} \Phi''(pa)\,(u_{r+p} + u_{r-p} - 2u_r). \tag{1.12}
$$

Thus, we obtain a linear homogeneous differential equation of second order. Its solution has the following form:

$$
u_r = A\exp\left[i\left(\omega t - \frac{2\pi}{\lambda}ra\right)\right]. \tag{1.13}
$$

This is the wave equation, see (1.1). Prove that it is actually the solution to the equation of motion and find the condition under which the solution occurs. Substituting (1.13) into (1.12) and using the Euler formula, we have

$$u_{r+p} + u_{r-p} - 2u_r = 2A \exp\left[i\left(\omega t - \frac{2\pi}{\lambda}ra\right)\right]\cos\left(\frac{2\pi}{\lambda}pa - 1\right)$$

$$= -4u_r \sin^2 \frac{2\pi}{\lambda}pa; \tag{1.14}$$

$$m\frac{d^2 u_r}{dt^2} = -mA\omega^2 \exp\left[i\left(\omega t - \frac{2\pi}{\lambda}ra\right)\right]. \tag{1.15}$$

As a result, we obtain

$$m\omega^2 = 2\sum_{p=1}^{s} \Phi''(pa)\left(1 - \cos\frac{2\pi}{\lambda}pa\right) \tag{1.16a}$$

or

$$m\omega^2 = 4\sum_{p=1}^{s} \Phi''(pa)\ \sin^2 \frac{\pi}{\lambda}pa. \tag{1.16b}$$

These are the conditions under which the equation of motion of the rth atom is described by the wave equation (1.13). The vibration frequency must satisfy relations (1.16). The same formulae imply that the squared cyclic frequency is a periodic function of the wave vector $q = 2\pi/\lambda$. If one adds the values $\pm 2\pi/\lambda$, $2 \times 2\pi/\lambda$, $3 \times 2\pi/\lambda,\ldots$ to the argument, then the left-hand side of (1.16) does not change.

We shall meet formulae of type (1.16) in Section 5, when analyzing the experimental data of measurements of frequencies of atomic vibrations in real crystals.

Two cases have a practical significance when we study the obtained q-dependence of ω:

(1) the chosen atom r interacts only with two nearest neighbours, i.e., with the atoms $r - 1$ and $r + 1$ (Fig. 1.3);

(2) a number s of atoms, whose distance from the atom r does not exceed the product sa, acts on the atom r.

Consider both of these cases. If only the nearest atoms interacts, then $p = 1$ and sum (1.16b) contains only one term. We immediately have

$$\omega = 2\sqrt{\Phi''/m}\ \sin(qa/2). \tag{1.17}$$

We have obtained a very important formula connecting the wave vector q and the angular frequency of the oscillatory mode ω. The

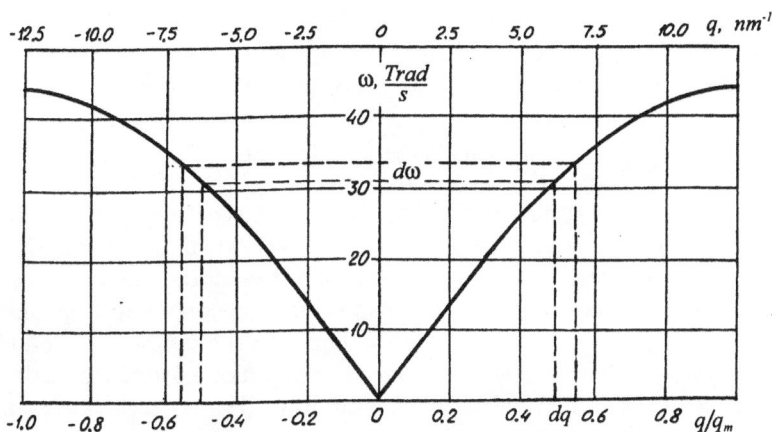

FIGURE 1.4. Dependence of the angular frequency of the vibrations of a linear one-atomic chain on the wave vector: the parameters of the atoms of nickel are used in the calculations, $a = 0.249$ nm; $m = 9.75 \times 10^{-26}$ kg; $q_m = 12.5$ nm^{-1}; $\omega_m = 44$ Trad/s. Approximate interaction of the nearest atoms. The dashed lines show small intervals of frequencies $d\omega$ and the corresponding wave vectors dq. The slope of the tangent $d\omega/dq$ at any point of the curve is numerically equal to the group velocity of propagation of the corresponding wave along the chain.

dependence of the cyclic frequency on the wave vector is called the *dispersion relation.*

The frequency of the elastic wave propagating in one-dimensional crystal depends on the values of the wave vector q or, and is the same, on the wave length λ. For relatively small q (long waves) this dependence is linear. With increasing q, the angular frequency increases according to the sinusoidal law achieving the maximal value at $q_m = \pi/a$; in this case, $\lambda = 2a$. The maximal frequency is $\omega_m = \sqrt{\Phi''/m}$. The dispersion relation, i.e., the curve of dependence of ω on q/q_m, is shown in Fig. 1.4.

Consider the speed of the wave propagation. It is known that the phase velocity is

$$v = \omega/q. \tag{1.18}$$

The group velocity $v_g = d\omega/dq$ is the velocity of motion of the envelope of the group of waves, which is localized in the space. Differentiating (1.17) by the wave vector, we obtain

$$v_g = a\sqrt{\Phi''/m}\,\cos(qa/2). \tag{1.19}$$

This is an interesting result! For waves of large length ($\lambda \gg 2a$), we have $\cos(qa/2) \approx 1$ and the waves have equal velocities; they increase as the force constant increases and the atom mass decreases. However, with increasing q, the group velocity decreases and vanishes at $q = \pi/a$, when the wave length is equal to the doubled period of the crystal lattice. Waves of smaller length cannot exist in the lattice. The wave of the minimal length is standing. The adjacent atoms in the standing wave move in antiphase.

Now, consider the second case, where the considered rth atom is affected by s atoms located in the distance sa; the affect of the more remote atoms can be neglected. From (1.16) we obtain

$$m\omega^2 = 2\sum_{p=1}^{s} \Phi''(pa)\left(1 - \cos\frac{2\pi}{\lambda}pa\right). \tag{1.20}$$

One can easily show that in this case the function $q(\omega)$ is multivalued: s values of the wave vector in the interval $[-\pi/a, +\pi/a]$ correspond to one value of the frequency. As an example, such curve is presented in Fig. 1.5.

Relations of form (1.20) play an important role when analyzing the nature of the interatomic bonding. From experimental curves of dependence $\omega^2(q)$, one can extract some information concerning the value of the force atomic constants of the crystal lattice. We will consider this in more detail.

Rewrite relation (1.20) in the following form:

$$\omega^2(q) = \frac{2}{m}\left\{\sum_{p=1}^{s}\Phi''(pa) - \sum_{p=1}^{s}\Phi''(pa)\cos qpa\right\}. \tag{1.21}$$

On the other hand, suppose that, from the experiment, we know the dependence of the squared angular frequency on the wave vector $\omega^2(q)$. This function is even, i.e., $\omega^2(q) = \omega^2(-q)$, and periodic. Expanding it into a finite Fourier series in cosines in the interval

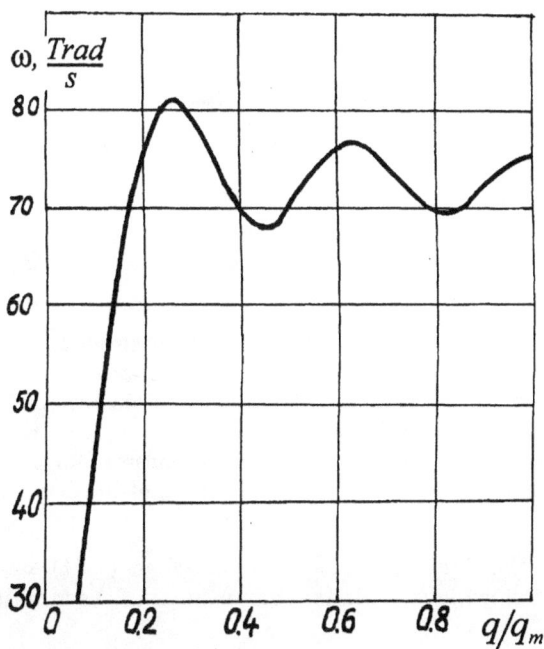

FIGURE 1.5. Dependence of ω on q/q_m: Any atom interacts with the five nearest atoms.

$[-\pi/a, \pi/a]$, we can write

$$\omega^2(q) = \frac{a_0}{2} + \sum_{p=1}^{s} a_p \cos qpa, \qquad (1.22)$$

where the Fourier coefficients a_p can be found by the formulae

$$a_p = \frac{a}{\pi} \int\limits_{-\pi/a}^{+\pi/a} \omega^2(q) \cos qpa \, dq. \qquad (1.23)$$

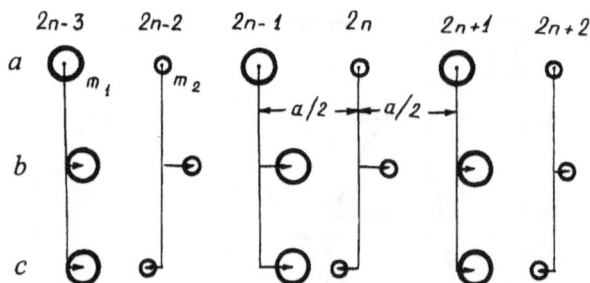

FIGURE 1.6. Vibrations of an infinite one-dimensional atomic chain with alternating atoms of different masses: a) — the atoms are in the equilibrium positions; b) — the atoms are displaced from the equilibrium positions, vibrate in one and the same phase; the vibrations correspond to the acoustic branch; c) — the atoms of different masses vibrate in antiphase; the vibrations correspond to the optical branch of the dispersion curve.

Comparing the right-hand sides of expressions (1.21) and (1.23), we obtain the following formula for computation of the atomic force constants:

$$\Phi''(pa) = -\frac{am}{2\pi} \int\limits_{-\pi/a}^{+\pi/a} \omega^2(q) \cos qpa \, dq. \qquad (1.24)$$

1.3. Chain with alternating atoms of different mass

The scheme of this atomic chain is shown in Fig. 1.6. All odd nodes are occupied by atoms of mass m_1 and even nodes are occupied by atoms of mass m_2. The distances between the adjacent atoms in equilibrium positions are the same and equal to $a/2$. Accordingly, the coordinates of the particles m_1 are $(2n + 1)a/2$ and those of the atoms m_2 are $(2n)a/2$. Assume that only the nearest particles interact and that the force constants are the same for the atoms of both kinds. Similarly to (1.12), we can write the equation of motion for the atoms of two kinds:

$$m_1\left(\frac{d^2 u_{2n+1}}{dt^2}\right) = \Phi''(u_{2n} + u_{2n+2} - 2u_{2n+1});$$

$$m_1\left(\frac{d^2 u_{2n}}{dt^2}\right) = \Phi''(u_{2n-1} + u_{2n+1} - 2u_{2n}). \qquad (1.25)$$

The solutions to these equations have the following form:

$$u_{2n+1} = A_1 \exp\left[i\left(\omega t - q(2n+1)a/2\right)\right];$$
$$u_{2n} = A_2 \exp\left[i\left(\omega t - q(2n)a/2\right)\right]. \tag{1.26}$$

The wave lengths and frequencies are equal for the atoms of both kinds, but the vibration amplitudes A_1 and A_2 are different. Substituting (1.26) into (1.25), we come to the system of equations:

$$A_1(m_1\omega^2 - 2\Phi'') + 2A_2\Phi'' \cos(qa/2) = 0;$$
$$A_2(m_2\omega^2 - 2\Phi'') + 2A_1\Phi'' \cos(qa/2) = 0. \tag{1.27}$$

In order to find ω, we set the determinant equal to zero:

$$\begin{vmatrix} m_1\omega^2 - 2\Phi'' & 2\Phi'' \cos(qa/2) \\ 2\Phi'' \cos(qa/2) & m_2\omega^2 - 2\Phi'' \end{vmatrix} = 0. \tag{1.28}$$

After some transformations, we obtain the following result:

$$\omega^2 = \frac{\Phi''}{m_1 m_2}\left[m_1 + m_2 \pm \sqrt{m_1^2 + m_2^2 + 2m_1 m_2 \cos qa}\right]. \tag{1.29}$$

Thus, in the case under consideration, two values of the angular frequency of the vibrations correspond to one value of the wave vector q. The dependence $\omega(q)$ is given in Fig. 1.7. The graphs contain two branches. The low-frequency branch of the dispersion curve is usually called acoustic, and the high-frequency one is called optical.

For long waves [minus sign in equation (1.29)], the frequency tends to zero as the wave vector decreases, and as it increases, the frequency first linearly increases, just as in the case of a one-atomic chain. The maximal value of the frequency in the acoustic branch is achieved at $q = \pi/a$, $q = -\pi/a$ and it is equal to

$$\omega_m = \sqrt{\frac{2\Phi''}{m}}. \tag{1.30}$$

For acoustic vibrations, the atoms vibrate in phase.

At $q = 0$, the optical branch has the maximum equal to

$$\omega = \sqrt{2\Phi''\left(\frac{1}{m_1} + \frac{1}{m_2}\right)}. \tag{1.31}$$

On further increasing the value of the wave vector, the frequency of the optical vibrations decreases according to the parabolic law and

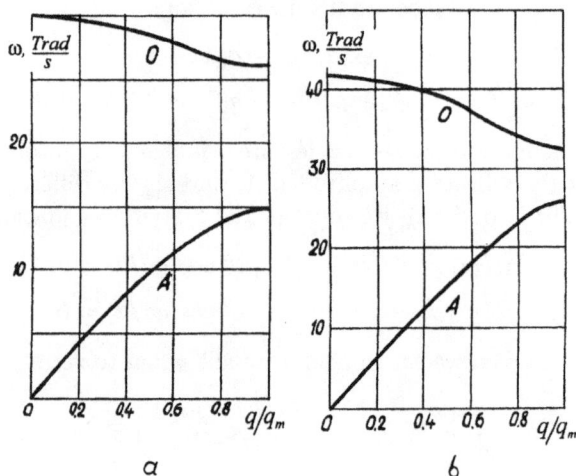

FIGURE 1.7. Dispersion curves for one-dimensional chains consisting of alternating atoms of two types: a) — atoms of AgCl, $m_1/m_2 = 3.04$; b) — atoms of NaCl, $m_1/m_2 = 0.65$. A — acoustic branch; O — optical branch.

at $q = \pm\pi/a$ we have

$$\omega_m = \sqrt{\frac{2\Phi''}{m_2}}. \tag{1.32}$$

In the region of optical vibrations, adjacent atoms deviate in antiphase.

It is useful to emphasize an interesting corollary of solution (1.29). For any value of $q = 2\pi/\lambda$, the sum of squared optical and acoustic frequencies is a constant:

$$\omega_A^2 + \omega_O^2 = 2\Phi''\frac{m_1 + m_2}{m_1 m_2} = \text{const}. \tag{1.33}$$

This assertion is the essence of the so-called rule of sum. We shall come back to it in Section 5.

1.4. Densities of vibrating states

In a crystal lattice, vibrations of different frequencies appear simultaneously, forming the vibrating spectrum. This spectrum is characterized by the function of vibration distribution by the frequency $g(\omega)$. The physical sense of this function is in the fact that, after multiplying by a small interval of frequencies $d\omega$, it gives the number of vibrations whose frequency lies in the interval $[\omega, \omega + d\omega]$. Sometimes, the quantity $g(\omega)\,d\omega$ is assumed to be equal to the part of the total number of vibrations belonging to the same frequency interval. Similarly, the number of vibrations in an interval $[\omega_1, \omega_2]$ is represented by the integral of the function of the distribution density computed in the appropriate limits:

$$N_{[\omega_1,\omega_2]} = \int_{\omega_1}^{\omega_2} g(\omega)\,d\omega. \tag{1.34}$$

Let us now see how one can pass from the dispersion curve $\omega(q)$ for a linear one-atomic chain to the distribution function $g = g(\omega)$. The interval of variation of the wave vector is $[-\pi/a, +\pi/a]$ (Fig. 1.4). The function $\omega(q)$ is periodic and is repeated beyond the limits of this interval, reflecting the periodicity of the crystal lattice. Thus, the length of the interval is $2\pi/a$. In this case, one can observe N values of the wave vector, just the number equal to the number of atoms in our model. Therefore, the number of wave vectors per interval unit is $N/(2\pi/a)$. And this is just the distribution function for the wave vectors:

$$G(q) = \frac{Na}{2\pi}. \tag{1.35}$$

Looking at Fig. 1.4, we see that the number of frequencies in the interval $d\omega$ is half the number of wave vectors in the interval dq. Hence,

$$g(\omega)\,d\omega = 2G(q)\,dq. \tag{1.36}$$

This implies the following equation for computation of the curve of frequencies distribution on the basis of the dispersion curve:

$$g(\omega) = \frac{2G(q)}{d\omega/dq}. \tag{1.37}$$

For the one-dimensional atomic chain with interaction only between the nearest atoms of equal mass, substituting formulae (1.17) and (1.35) into (1.37), we obtain

$$g(\omega) = \frac{2N}{\pi}\left(\frac{1}{\sqrt{\omega_m^2 - \omega^2}}\right). \tag{1.38}$$

The curve of the density of the vibrations states for this case (Fig. 1.8) begins for $\omega = 0$ at the ordinate $2N/\pi\omega_m$ and infinitely increases as

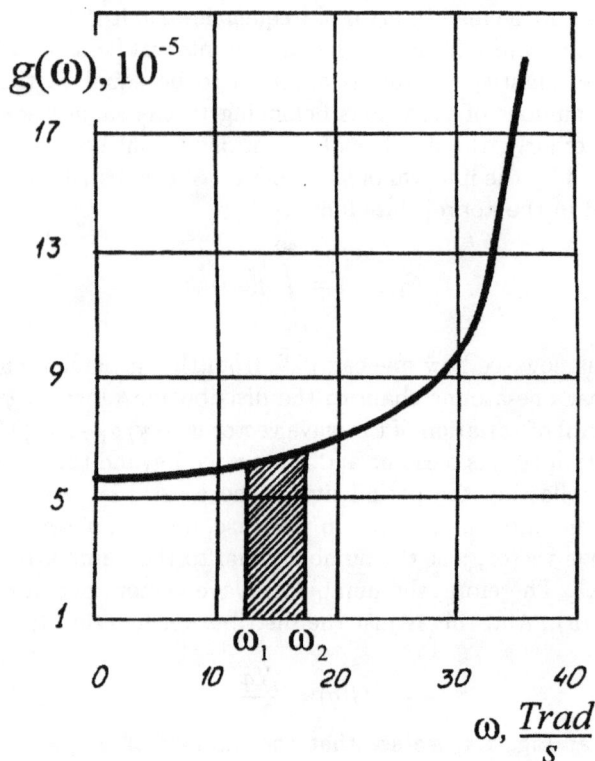

FIGURE 1.8. The distribution function for a one-dimensional chain with respect to frequencies. The area of the hatched figure illustrates the number of vibrations, whose frequencies lie in the interval $[\omega_1, \omega_2]$. The number of high-frequency modes per unit interval $d\omega$ is larger than those of low-frequency ones.

$\omega \to \omega_m$. However, the total number of vibrations is, of course, equal to the total number of atoms in the considered one-dimensional chain:

$$\int_O^{\omega_m} g(\omega) \, d\omega = N. \tag{1.39}$$

A typical spectral distribution of frequencies is given by an atomic chain with alternating different masses. The spectrum of such a chain

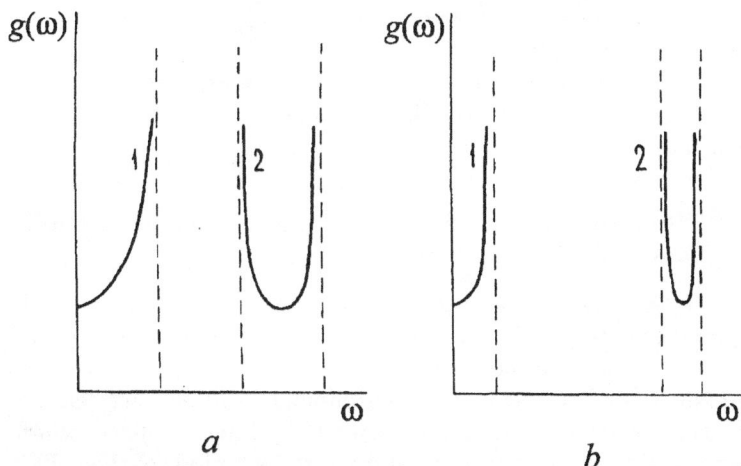

FIGURE 1.9. The frequency distribution function (vibration spectrum) for a two-atomic chain: a) — the ratio of the masses is equal to 4; b) — the ratio of the masses is equal to 9. 1 — acoustic frequencies; 2 — optical frequencies. Both the domains are separated from each other by the area of the forbidden frequencies. At the ends of the intervals shown by dashed lines, the curves go to infinity.

consists of two branches (Fig. 1.9). Both the branches are considerably disjoint and the form of the spectrum essentially depends on the ratio of the masses. The distance between the branches along the axis of frequencies is equal to $\left(\sqrt{2\Phi''/m_2} - \sqrt{2\Phi''/m_1}\right)$ and increases with increasing of the ratio of the atomic masses. The spectrum of the low-frequency, acoustic branch is similar to the spectrum of the one-atomic lattice shown in Fig. 1.8. It contains N eigen frequencies. The optical part of the spectrum is rather narrow, especially for large values of the ratio m_1/m_2. Practically all optical frequencies are concentrated in one domain, the spectrum is almost monochromatic. The optical part of the spectrum also contains N eigen frequencies.

Thus, two kinds of the atoms in one-dimensional lattice imply the appearance of two domains in the spectral distribution of the frequencies $g(\omega)$. In essence, this is also typical for real three-dimensional crystals. However, infinitely high peaks are appropriate only for the model one-dimensional chain.

The real vibrating spectra of solids will be considered in Section 5. Note that, in a three-dimensional crystal, the total number of vibrations is $3N$ due to the growth of the degrees of freedom of atoms. One third of the oscillatory modes is longitudinal, and two thirds are transverse.

1.5. Three-dimensional crystal. Atomic force constants. Concept of the dynamic matrix

Now, we consider the general case of a three-dimensional crystal having an elementary cell with a basis. The cell contains n atoms. For instance, for a cubic body-centered crystal lattice, we have $n = 2$. The position of the elementary cell number l in the space relative to the origin is determined by the vector $\mathbf{r}(l) = u\mathbf{a} + v\mathbf{b} + w\mathbf{c}$, where u, v, and w are integers and $\mathbf{a}$, $\mathbf{b}$, and $\mathbf{c}$ are the periods of the cell. The position of the chosen atom number k in the cell is determined by the vector $\mathbf{r}(k)$ (Fig. 1.10). The quantity k can take the values from 1 to n. It is convenient to introduce the notation $\mathbf{r}(lk)$, where the first letter in parenthesis determines the position of the cell, and the second letter determines the location of the atom in this cell. Thus, $\mathbf{r}(lk) = \mathbf{r}(l) + \mathbf{r}(k)$.

In the process of vibration, the atoms deviate from the equilibrium positions. The displacement vector depends on the atom coordinates and it is convenient to denote it similarly $\mathbf{u}(lk)$. To take account of the direction of the displacements, their components are denoted by α and β. Correspondingly to three coordinate axes, $\alpha, \beta = 1, 2$, or 3. The displacement of the atom (lk) in the direction α is denoted $\mathbf{u}_\alpha(lk)$. Similarly, another atom $(l'k')$ deviates, in general case, on a vector $\mathbf{u}_\beta(l'k')$.

The kinetic energy of a vibrating crystal is expressed by the following sum:

$$T = \frac{1}{2} \sum_{l,k,\alpha} m\dot{u}^2(lk). \tag{1.40}$$

The summation is over all atoms and all directions of displacements.

The potential energy of the crystal Φ is a function of mutual positions of all atoms. Let us expand the expression for the potential energy of the three-dimensional lattice into the Taylor series in the powers of the atomic displacements. Supposing that the displacements are small, we restrict ourselves by the powers of displacements

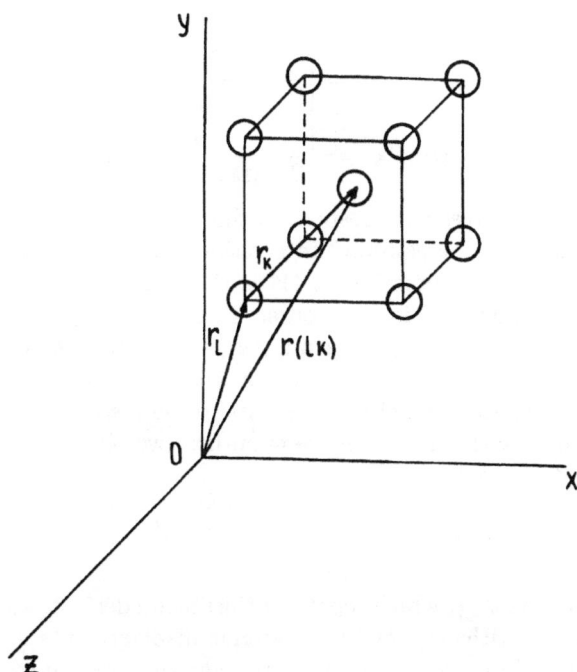

FIGURE 1.10. The vectors determining the atom position in an imprimitive elementary cell.

not exceeding the second one. This is the so-called harmonic approximation. We obtain

$$\Phi = \Phi_0 + \sum_{l,k,\alpha} \Phi_\alpha(lk)u_\alpha(lk)$$

$$+ \frac{1}{2} \sum_{\substack{l,k,\alpha \\ l',k',\beta}} \Phi_{\alpha\beta}(lk; l'k')\, u_\alpha(lk)u_\beta(l'k'). \quad (1.41)$$

The quantity Φ_0 is the balanced potential energy of the crystal; it does not depend on the displacements and only changes the reference point of the potential energy. In equation (1.41), we use the following notation;

$$\Phi_\alpha(lk) = \left[\frac{\partial \Phi}{\partial u_\alpha(lk)}\right]_0; \qquad (1.42)$$

$$\Phi_{\alpha\beta}(lk; l'k') = \left[\frac{\partial^2 \Phi}{\partial u_\alpha(lk)\partial u_\beta(l'k')}\right]_0, \qquad (1.43)$$

where the derivatives with respect to the displacements of the atoms are determined in the equilibrium positions. The second term on the right-hand side of expression (1.41) is the force acting on the atom in the equilibrium position; of course, this term is equal to zero.

The force acting onto the atom (lk) is equal to minus gradient of the potential energy of its interaction with all other atoms. Differentiating expression (1.41) with respect to the displacements $u_\alpha(lk)$, in accordance with the second Newton law, we obtain the following equation of motion:

$$m_k \ddot{u}_\alpha(lk) = -\frac{1}{2} \sum_{l',k',\beta} \Phi_{\alpha\beta}(lk; l'k')\, u_\beta(l'k'). \qquad (1.44)$$

The coefficients $\Phi_{\alpha\beta}$, which represent the second derivatives of the potential energy with respect to the atomic displacements, determined at the equilibrium points are called the *atomic force constants*. They have an explicit physical sense. The coefficient denoted $\Phi_{\alpha\beta}(lk; l'k')$ is equal to minus force which acts on the atom (lk) in the direction α, when the other atom $(l'k')$ deviates per unit distance in the β-direction assuming that all other atoms stay put at their equilibrium positions. The atomic force constants satisfy the symmetry conditions: it makes no difference whether the atom $(l'k')$ deviates and the atom (lk) remains at rest, or, conversely, the atom (lk) deviates and $(l'k')$ is fixed. The force constants depend on the difference of l and l', not on these quantities themselves. Therefore,

$$\Phi_{\alpha\beta}(l\,k;\,l'\,k') \;=\; \Phi_{\beta\alpha}(l'\,k';\,l\,k) \;=\; \Phi_{\alpha\beta}(0\,k;\,l - l'\,k'). \qquad (1.45)$$

Any atomic force constant reflects the linear relation between the displacement of the atom and the value of the restoring force. For instance,

$$F_\alpha = \frac{\partial^2 \Phi}{\partial u_\alpha \partial u_\beta} u_\beta. \qquad (1.46)$$

In the notation $\Phi_{\alpha\beta}$, the first subscript is the component of the force, and the second one is the component of displacement of the atom.

When two atoms interact, six different force constants can appear: Φ_{11}, Φ_{12}, Φ_{13}, Φ_{22}, Φ_{23}, and Φ_{33}. Due to the symmetry of the crystal lattice, the number of independent atomic force constants is less than six. For instance, in a cubic body-centred lattice, when the atom located at the centre of the elementary cell is displaced by the unit distance, the following forces act on the atom located at the origin: $\Phi_{11} = \Phi_{22} = \Phi_{33}$; $\Phi_{12} = \Phi_{13} = \Phi_{23}$. Thus, there are only two independent force constants in the first coordinate sphere.

The motion equations obtained (1.44) for atoms in a three-dimensional crystal are, principally, analogous to equations (1.12) obtained before for the one-dimensional chain. The solution to the motion equations has the following form:

$$u_\alpha(lk) = u_\alpha^0 \exp\left[i\left(\omega t - \mathbf{q}\mathbf{r}(lk)\right)\right], \qquad (1.47)$$

where u_α^0 is the amplitude. Similarly to the displacement of the atom $(l'k')$, we have

$$u_\beta(l'k') = u_\beta^0 \exp\left[i\left(\omega t - \mathbf{q}\mathbf{r}(l'k')\right)\right]. \qquad (1.48)$$

Both the latter equations describe the wave displacement of the atoms [see (1.1)]. Substituting (1.47) and (1.48) into equation (1.44), we obtain

$$\omega^2 u_\alpha^0 = \frac{1}{m_k} \sum_{l',k',\beta} \Phi_{\alpha\beta}(lk; l'k')\, u_\beta^0 \exp\left\{i q\left[\mathbf{r}(lk) - \mathbf{r}(l'k')\right]\right\}. \qquad (1.49)$$

Introduce the following notation:

$$D_{\alpha,\beta}(qkk') = \frac{1}{m_k} \sum_{l',k',\beta} \Phi_{\alpha\beta}(lk; l'k') \exp\left\{i q\left[\mathbf{r}(lk) - \mathbf{r}(l'k')\right]\right\}, \qquad (1.50)$$

where the sum is taken over all components of the vector $\mathbf{r}(l'k')$, which are equal to $\mathbf{a}'$, $\mathbf{b}'$, and $\mathbf{c}'$. With due account of this notation, the solution to the motion equation is obtained in the form

$$\omega^2 u_\alpha^0 = D_{\alpha\beta}(qkk') u_\beta^0. \qquad (1.51)$$

The quantities $D_{\alpha\beta}(qkk')$ are the elements of the dynamical matrix; definition (1.50) implies that they are the Fourier transforms of the force constants.

For a given value of the wave vector q, the angular frequency of the vibrations ω is defined by the following characteristic equation:

$$\left| D_{\alpha\beta}(qkk') - \omega^2 \delta_{kk'} \delta_{\alpha\beta} \right| = 0, \qquad (1.52)$$

where $\delta_{kk'}$ and $\delta_{\alpha\beta}$ are the Kroneker symbols:

$$\delta_{ij} = \begin{cases} 1, & i = j; \\ 0, & i \neq j. \end{cases}$$

Considering the pairs of the indices α, β in the following order $(1,1)$, $(2,1)$, $(3,1)$, $(1,2)$, $(2,2)$, $(3,2)$, $\ldots$, $(1,n)$, $(2,n)$, $(3,n)$, we obtain the dynamic matrix in the following form (the commas between the indices are omitted):

$$\begin{matrix} D_{11}(q11) & D_{12}(q11) & D_{13}(q11) & D_{11}(q12) & \cdots & D_{13}(q1n) \\ D_{21}(q11) & D_{22}(q11) & D_{23}(q11) & D_{21}(q12) & \cdots & D_{23}(q1n) \\ D_{31}(q11) & D_{32}(q11) & D_{33}(q11) & D_{31}(q12) & \cdots & D_{33}(q1n) \\ \\ \cdots & \cdots & \cdots & \cdots & \ddots & \cdots \\ D_{31}(qn1) & D_{32}(qn1) & D_{33}(qn1) & D_{31}(qn2) & \cdots & D_{33}(qnn) \end{matrix}$$

$$(1.53)$$

The dynamic matrix contains $3n$ rows, each row contains $3n$ elements. For example, the element $D_{12}(q13)$ denotes the Fourier transform of the force acting on the atom number 1 in the direction Ox, if the atom number 3 in the elementary cell is displaced on the unit distance in the direction Oy. The left-hand side of equation (1.52) is a brief notation of the characteristic determinant of matrix (1.53). Equation (1.52) is an equation of degree $3n$ in ω^2. This equation determines $3n$ frequencies. For each of these frequencies, we have a separate solution in form (1.47).

Thus, for any value of the wave vector q, we obtain $3n$ linear equations with $3n$ unknown quantities ω^2.

Equations of types (1.50) and (1.52) are usually used for calculating the spectra $g(\omega)$ of three-dimensional crystals. Here, as a rule, model crystals consisting of a number of atoms are considered. The model of the interatomic forces and the values of the force constants should be given. The spectra calculated in this way can be compared with those obtained in experiments, and conclusions concerning the correctness of accepted assumptions can be made on the base of this comparison. For example, the curves calculated for copper and silver are in good agreement with experimental dependencies.

1.6. Amplitudes of the atomic vibrations

The vibrating displacements of the atoms from the equilibrium position occur in different directions. The arithmetic mean atomic displacement is equal to zero, because all directions of displacements of atoms from the equilibrium position in a crystal lattice are equiprobable. Introducing the mean-square amplitude $\sqrt{\overline{u^2}}$, one can get rid of negative values of displacements. The displacement vector **u** can be decomposed along the three coordinate axes. Each of $3N$ waves of the lattice has its frequency ω_i and its amplitude u_i. The value of the mean-square amplitude can be calculated, if the frequency distribution function $g_i(\omega)$ is known. According to the definition of the mean value of a function, we have

$$\overline{u^2} = \frac{\sum_{i=1}^{N} \sum_{\alpha=1}^{3} u_{i\alpha}^2(\omega)\, g_i(\omega)}{\sum_{i=1}^{N} g_i(\omega)}, \qquad (1.54)$$

where the sum is taken over all waves of the crystal lattice and three directions of displacements.

Choose an independent Cartesian coordinate system in the lattice. Then, the squared amplitude of the atomic displacements in some direction is

$$\overline{u_{i\alpha}^2} = \overline{u_{ix}^2} + \overline{u_{iy}^2} + \overline{u_{iz}^2}. \qquad (1.55)$$

Since the heat displacements along all axes are equiprobable, the terms in the right-hand side are equal to each other. Therefore, the mean-square amplitude of the ith oscillatory mode along one of the axes is

$$\overline{u_{ix}^2} = \frac{1}{3}\overline{u_{i\alpha}^2}. \qquad (1.56)$$

It is interesting to find the dependence between the frequency and the amplitude of the oscillatory mode [8]. It has been said that, in a three-dimensional simple crystal lattice, N wave propagate in three independent directions. The displacement of the nth node can be written in the following way:

$$u_n = \sum_{i=1}^{N} \sum_{\alpha=1}^{3} u_{i\alpha} \sin\left(\omega_{i\alpha} t - q_{i\alpha} x\right), \qquad (1.57)$$

where $u_{i\alpha}$ is the amplitude of the ith oscillatory mode in the direction α, $q_{i\alpha}$ is the corresponding wave vector. The sum is taken over N values of i and three values of α.

The total kinetic energy of the crystal lattice consisting of equal atoms of mass m can be found by the obvious formula:

$$E_{\text{kin}} = \frac{1}{2} \sum_{n=1}^{N} m \dot{u}_n^2. \tag{1.58}$$

Differentiating sum (1.57) with respect to time, we obtain the following expression for the velocity of the node n:

$$\dot{u}_n = \sum_{i=1}^{N} \sum_{\alpha=1}^{3} u_{i\alpha} \omega_{i\alpha} \cos(\omega_{i\alpha} t - q_{i\alpha} x). \tag{1.59}$$

The square of the mean-square value of the velocity can be determined, by squaring the right-hand side of this expression and averaging every term of the sum. It should be taken into account that the mean value of the squared cosine is equal to $1/2$ and the mean value of the product of two cosines is equal to zero, since the waves are independent. Thus, we have

$$\dot{u}_n^2 = \frac{1}{2} \sum_{i=1}^{N} \sum_{\alpha=1}^{3} u_{i\alpha}^2 \omega_{i\alpha}^2. \tag{1.60}$$

The mean value of the total energy $\overline{E}$ is equal to the doubled value of E_{kin}. Therefore, from (1.58) and (1.60) we obtain $\overline{E_{i\alpha}} = \frac{1}{2} N m \omega_{i\alpha}^2 \overline{u_{i\alpha}^2}$ or, finally,

$$\sqrt{\overline{u_{i\alpha}^2}} = \frac{\sqrt{2\overline{E_{i\alpha}}/m}}{\omega_{i\alpha}}. \tag{1.61}$$

This implies two important conclusions.

The mean-square amplitude of the oscillatory mode is inversely proportional to the frequency. Smaller amplitudes must correspond to the vibrations of higher frequencies.

One can see from the same equation (1.61) that larger amplitudes are typical for atoms of a smaller mass than those for more massive atoms.

According to the Planck formula, the mean energy of the quantum oscillator

$$\overline{E_{i\alpha}} = \left[\frac{\hbar \omega_{i\alpha}}{\exp\left(\frac{\hbar \omega_{i\alpha}}{kT}\right) - 1} + \frac{\hbar \omega_{i\alpha}}{2} \right], \tag{1.62}$$

where $\hbar = h/2\pi$; h is the Planck constant; T is the temperature.

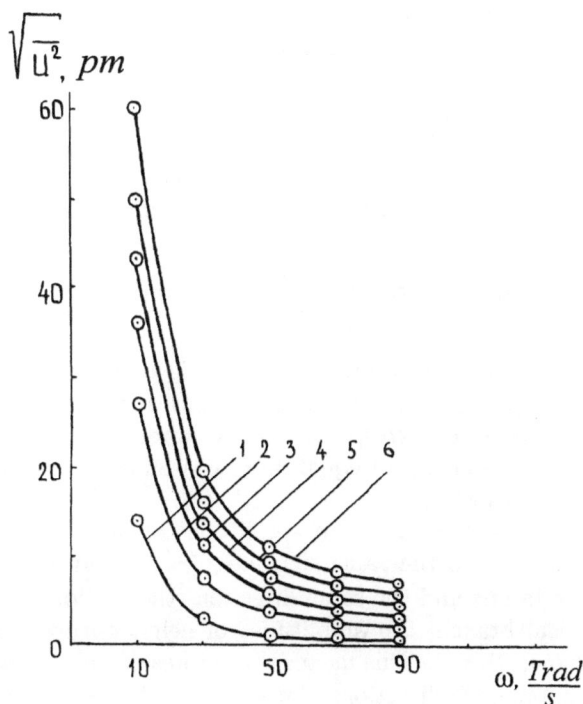

FIGURE 1.11. Dependence of the mean-square amplitude of the atomic vibrations on the temperature and frequency. The curves are calculated by formulae (1.61) and (1.62) for atoms of nickel: temperatures: 1 — 100; 2 — 300; 3 — 500; 4 — 700; 5 — 900; 6 — 1300 K.

This formula can be used for calculation of the dependence of the mean-square amplitude on the frequency according to relation (1.61). Such curves are shown in Fig. 1.11.

Suppose that the temperature is sufficiently high, so that the condition $\hbar\omega_{i\alpha} \ll kT$ holds. Then, one can neglect the second term in the right-hand side of (1.62) and expand the exponent into a series restricting itself by two terms. As a result, the formula takes the form $\overline{E_{i\alpha}} = kT$. Hence, under this condition we have

$$\overline{u_{i\alpha}^2} = \frac{2kT}{mN\omega_{i\alpha}^2}. \tag{1.63}$$

We obtain that the square of the mean-square amplitude in the domain of relatively high temperatures is proportional to the absolute temperature.

Now, consider the amplitudes of vibrations of a two-atomic crystal on the example of the one-dimensional chain (Fig. 1.6). From equations (1.27) and (1.29) one can obtain the following ratio A_1/A_2 of the amplitudes of the optical and acoustic vibrations, respectively, $(m_1 > m_2)$:

$$\frac{A_1}{A_2} = \frac{(m_1 - m_2) \mp \sqrt{m_1^2 + m_2^2 + 2m_1 m_2 \cos 2qa}}{2m_1 \cos qa}. \tag{1.64}$$

The minus sign at the square root in the last formula corresponds to the plus sign in (1.29); hence, to the optical branch of the curve $\omega(q)$. The plus sign corresponds to the lower, acoustic branch of this curve. The analysis of equation (1.64) allows one to make the following conclusions [3].

In the acoustic branch, the amplitudes of neighbouring atoms have always one and the same sign, i.e., $\frac{A_1}{A_2} > 0$. In this case, the atoms move in one and the same direction (the motion "in phase"). In the optical branch, the amplitudes of neighbouring atoms have different signs, $\frac{A_1}{A_2} < 0$. This means that atoms of the masses m_1 and m_2 move in opposite directions, "in antiphase".

For $qa = 0$, atoms of both kinds in the acoustic branch have equal amplitudes, fulfilling in vibrations the pure translation of the lattice. In the optical branch, the atoms deviate in opposite directions, and the ratio of the amplitudes is inversely proportional to the ratio of the masses of the atoms. When the value qa increases from 0 to π, the amplitudes ratio in the acoustic branch grows, and in the optical branch tends to zero and that means that more massive atoms stay put.

1.7. On the Debye model

Debye made the fundamental assumption, according to which the normal mode of the vibrations has the mean energy of the Planck vibration, see (1.62). The Debye model is described in a number of publications, [5, 8, 12]. Its detailed consideration and use are not covered in this review. However, in many publications, results of the measurements of the characteristic (Debye) temperature for a number of materials are presented. Therefore, we sketch the peculiarities of the Debye model, developed mainly in order to explain the temperature dependence of the specific heats of solids; classical physics cannot explain this dependence.

According to Debye, there exists a maximal frequency $\nu_m = \nu_D$ of the lattice vibrations, which is determined by its discreteness. The characteristic Debye temperature θ corresponds to this frequency:

$$h\nu_m = k\theta. \tag{1.65}$$

The frequency distribution function is given by the following equation:

$$g(\nu) = \begin{cases} \dfrac{9N\nu^2}{\nu_D^2} & \text{for} \quad 0 \le \nu \le \nu_D; \\ 0 & \text{for} \quad \nu > \nu_D. \end{cases} \tag{1.66}$$

$$\int_0^{\nu_D} g(\nu)\, d\nu = 3N. \tag{1.67}$$

In conclusion, we give the formula connecting the Debye temperature with the square of the mean-square displacement [12]:

$$\overline{u^2} = \frac{9h^2 T}{4\pi^2 m k \theta^2} \big[\Phi(x) + 0.25x \big], \tag{1.68}$$

where $x = \theta/T$, and the Debye function is written in the brackets on the right-hand side. For $x < 1$ (relatively high temperatures) its value is close to one.

2. EXPERIMENTAL METHODS FOR MEASUREMENTS OF THE MEAN-SQUARE AMPLITUDES OF THE ATOMIC DISPLACEMENTS AND CHARACTERISTICS OF THE FREQUENCY SPECTRUM

2.1. Influence of the dynamic atomic displacements on the scattering of X-rays

The wave length of electromagnetic radiation in the X-ray range has the same order as the interatomic distances in solids. When a crystal is irradiated, the X-rays excite the electrons in the atom shells forcing them to vibrate with the frequency equal to the vibration frequency of the vector of the electric field intensity of the original wave. The rays scattered by many atoms interact with each other. The interference occurs and the resulting electromagnetic vibration propagates in certain selective directions. The result of the interference depends on the distance between the atoms. The heat vibrating motion of the atoms essentially influences to the interference picture.

First, consider the simplest case of interference of electromagnetic waves of X-ray range by two atoms (Fig. 2.1). The first atom 0 is at the origin; the location of the second atom N is determined by the radius-vector $\mathbf{r}$. The direction of the incident rays is given by the vector $\mathbf{s_0}$; the direction of the scattered rays, by the vector $\mathbf{s}$. The lengths of both the vectors are equal to one, i.e., $|\mathbf{s_0}| = |\mathbf{s}| = 1$. (Sometimes, the authors choose $|\mathbf{s_0}| = |\mathbf{s}| = 2\pi/\lambda$. This choice, firstly, simplifies the formulae; and secondly, makes them nearer to the generally accepted form equations describing the neutron diffraction on elastic vibrations of a lattice.)

a

b

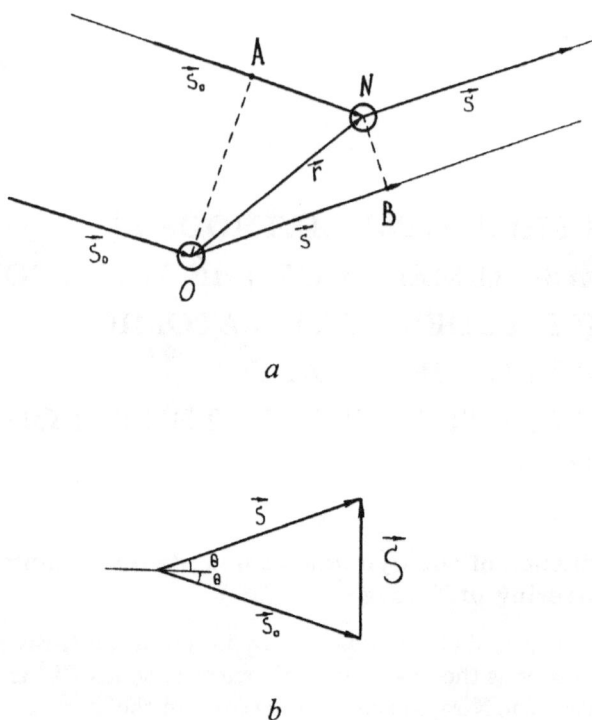

FIGURE 2.1. The scheme of X-ray scattering by two atoms: a — path of rays; 0 — the atom at the origin; N — the atom whose position is determined by the radius-vector **r**; b — the scheme of the location of the vectors; the trace of "reflecting" crystalline plane is shown.

Let us return to Fig. 2.1. The difference of the path lengths of the two rays — scattered and incident – is equal to $d = OB - OA$. It can be found as the difference of the scalar products

$$d = (\mathbf{rs}) - (\mathbf{rs_0}) = \mathbf{r}(\mathbf{s} - \mathbf{s_0}) = (\mathbf{rS}), \qquad (2.1)$$

where **S** is the difference of the diffracted and incident vectors. One can see from Fig. 2.1b that its length is $|\mathbf{S}| = 2\sin\theta$, where θ is the angle of "reflection" from the crystalline planes. The difference of the phases of the rays scattered by two atoms can be obtained by multiplying the difference of the pathes by $2\pi/\lambda$ (in accordance with the proportion $2\pi : \lambda = \varphi : d$):

$$\varphi = \frac{2\pi}{\lambda}(\mathbf{rS}) = (\mathbf{rQ}), \qquad (2.2)$$

where $\mathbf{Q} = 2\pi\mathbf{S}/\lambda$ is the diffraction vector. Its length is equal to $Q = (4\pi/\lambda)\sin\theta$. This implies that if the wave scattered by the zero atom at a point of the space induces the displacement $Y = Y_0 \exp i\omega t$, then the displacement induced by the wave scattered by the atom N is equal to

$$Y = Y_0 \exp\{i[\omega t - (\mathbf{rQ})]\}. \qquad (2.3)$$

The amplitude of the total wave Y_0 depends on the scattering power of this atom characterized by the factor of the atomic scattering f.

The result obtained (2.3) should be summarized over all scattering nodes. In the case of a primitive crystal lattice (one atom at the cell) and ideal location of the atoms, the radius-vector of any of them is determined by three integers u, v, and w:

$$\mathbf{r} = u\mathbf{a} + v\mathbf{b} + w\mathbf{c}, \qquad (2.4)$$

where $\mathbf{a}$, $\mathbf{b}$, and $\mathbf{c}$ are parameters of the lattice. The resulting displacement in the wave scattered by a crystal containing N atoms can be written in the following way:

$$Y = Y_0 \exp(i\omega t) \sum_{n=1}^{N} \exp[-i(\mathbf{r}_n\mathbf{Q})]. \qquad (2.5)$$

In order to obtain the intensity of the scattering by a primitive lattice, we have to multiply Y by its complex-conjugate Y^*. As a result, we obtain the double sum

$$Y = Y_0 \sum_{m=1}^{N} \sum_{n=1}^{N} \exp\{i[(\mathbf{r}_m - \mathbf{r}_n)\mathbf{Q}]\}. \qquad (2.6)$$

The sum in the last expression should be taken over all nodes m and n from 1 to the total number of scattering nodes of the lattice N; in this case, the sums contain the unit terms for which $m = n$.

Equation (2.6) has a totally general character. We emphasize that the intensity of the scattered wave is determined not by the radius-vectors of the atoms, but by the vectors of distances between all pairs of the atoms.

Suppose that the scattering nodes are displaced from the regular positions by the displacement vectors $\mathbf{u}_1, \mathbf{u}_2, \ldots, \mathbf{u}_m, \mathbf{u}_n$. Then, sums of the forms $\mathbf{r}_1 + \mathbf{u}_1$, $\mathbf{r}_2 + \mathbf{u}_2$, ... must be introduced into formula (2.5). The scattering intensity can be written as

$$I = Y_0^2 \sum_{n=1}^{N} \sum_{m=1}^{N} \exp\{-i[(\mathbf{r}_n - \mathbf{r}_m)\mathbf{Q}]\} \exp\{-i[(\mathbf{u}_n - \mathbf{u}_m)\mathbf{Q}]\}. \quad (2.7)$$

The heat motion of the atoms, affecting the displacements, results in the change of the conditions of the X-rays diffraction. The first factor in the sums in (2.7) is determined only by the crystal structure and do not depend on time. The vibration frequency of atoms is of the order 10^{12} Hz, whereas the typical frequency of X-rays exceeds it in 10^6 times. Thus, the X-ray "meets" as though an stationary atoms distribution. The difference of displacements of every pair of atoms must be averaged in time. We can take into account the fact that the mean value of the difference $\overline{\mathbf{u}_n - \mathbf{u}_m} = \overline{\Delta\mathbf{u}}$ at any fixed instant is equivalent to the mean value of the difference of displacements in time for any chosen atomic pair. If the displacements are small, then the exponent can be expanded into a series keeping only three terms

$$\overline{\exp[-i(\Delta\mathbf{u}\mathbf{Q})]} = 1 - i\overline{(\Delta\mathbf{u}\mathbf{Q})} - 0.5\overline{(\Delta\mathbf{u}\mathbf{Q})^2} + \ldots \quad (2.8)$$

The second term on the right-hand side of (2.8) is equal to zero, because positive and negative values of the difference of displacements parallel to any direction are equiprobable. Then, from (2.8) we obtain

$$1 - 0.5\overline{[(\mathbf{u}_n - \mathbf{u}_m)\mathbf{Q}]}^2 = 1 - 0.5\left[\overline{\mathbf{u}_n^2\mathbf{Q}^2} - \overline{2\mathbf{u}_n\mathbf{u}_m\mathbf{Q}} + \overline{\mathbf{u}_m^2\mathbf{Q}^2}\right]. \quad (2.9)$$

The vibrations of different nodes are independent and have one and the same mean energy. This means that $\overline{\mathbf{u}_n\mathbf{u}_m} = 0$ and $\overline{\mathbf{u}_n^2} = \overline{\mathbf{u}_m^2}$. Considering again the exponential dependence of the displacements and omitting the indices, we obtain that the mean value of the second factor in formula (2.7) is equal to

$$1 - \overline{u^2}Q^2 = \exp(-\overline{u^2}Q^2) = \exp\{-\overline{u^2}[16\pi^2(\sin^2\theta/\lambda^2)]\}. \quad (2.10)$$

Clearly, $\sqrt{\overline{u^2}}$ is the displacement parallel to the projection of $\mathbf{u}$ onto $\mathbf{Q}$. In the Cartesian coordinate system, $\overline{u^2} = \overline{u_x^2} + \overline{u_y^2} + \overline{u_y^2} = 3\overline{u_x^2}$, since the displacements are equiprobable.

Thus, the intensity of the scattering of the X-rays by a group of atoms subjected to independent heat vibrations is weakened due to the factor $\exp(-2M)$, where

$$-2M = -\frac{16}{3}\pi^2\overline{u^2}\frac{\sin^2\theta}{\lambda^2}. \quad (2.11)$$

For an inprimitive cell, we use the notation introduced above $\mathbf{r}(lk)$ for the radius-vector. Then, equation (2.5) takes the form

$$Y = Y_0 \exp(i\omega t) \sum_{l=1}^{N} \sum_{k=1}^{n} \exp\{-i[\mathbf{r}(lk)\mathbf{Q}]\}, \qquad (2.12)$$

where N is the number of cells and n is the number of atoms in an elementary cell. In such crystal lattice with a basis, the Debye–Waller factor $\exp(-2M)$ is present in the expression for the structural factor. The scattering intensity proportional to the square of the structural amplitude is written in the following way:

$$I = I_0 \left\{ \sum_{k=1}^{n} f_k \exp(-M_k) \exp\left[-2\pi i(hx_k + ky_k + lz_k)\right] \right\}^2 pL,$$
$$(2.13)$$

where I_0 is the integral intensity if the incident beam; f_k is the factor of the atomic scattering of the kth atom; x_k, y_k, and z_k are the coordinates of the atoms in an elementary cell in units of period; p is the repetition factor of the reflecting plane with the Miller indices (hkl); L is the product of the Lorentz factor and the angular one.

If the elementary cell contains atoms of only one sort, then the formula (2.13) can be simplified, since the atomic factor and the Debye–Waller can be factored out from the sum.

In the general case, an elementary cell of two- and three-component compounds contains different atoms. Obviously, the value of the mean-square displacements is different for atoms of different sorts. To measure them, some special methods should be used and these are considered.

There are two versions of experimental methods for the determination of the mean-square atomic displacements in one-component solid solutions. G.V. Kurdyumov and his coworkers used the measurements of one and the same interference line, which were made at two temperatures (of liquid nitrogen and room temperature) [9, 10]. Here, all intensity factors, except the Debye–Waller one, are practically equal. Taking the natural logarithm of the intensity ratio for the two temperatures, in accordance with (2.13) and with due account of (2.11), we obtain

$$\ln(I'/I) = -\frac{16}{3}\pi^2 \left(\overline{u^{2\prime}} - \overline{u^2}\right) \frac{\sin^2\theta}{\lambda^2}, \qquad (2.14)$$

where the primed quantities refer to the high temperature and those without primes, to the lower one $(\sin^2 \theta' \cong \sin^2 \theta)$. This implies that, in essence, this method allows one to determine the difference $\overline{\Delta u^2}$ of the mean-square displacements of atoms for two temperatures.

In our experiments, the specimens of metals and alloys were studied at a range of high temperatures as well as at room temperature. The technique described requires special X-ray cameras which provide the thermostatation of the specimens during the X-ray investigation.

The second version of the method allows one to find the mean-square displacements for a given temperature analyzing the integral intensity of several interferences. After taking the logarithm, expression (2.13) can be reduced to the following form:

$$\ln(I/pL|F|^2) = \ln(I_0 C) - \frac{16}{3}\pi^2 \overline{u^2} \frac{\sin^2 \theta}{\lambda^2}, \qquad (2.15)$$

where C is a constant. The experimental data should be corrected due to the diffuse scattering [11]. Then, constructing the graph in the coordinates "$\ln(I/pL.|F|^2)$ —$\sin^2 \theta/\lambda^2$", we obtain a linear dependence and calculate $\overline{u^2}$ by the slope of this straight line with respect to the x-axis.

For a two-atomic elementary cell, the measurements of the heat displacements are based on the general formula (2.13). For instance, a basis of an elementary cell of a crystal lattice of the caesium chloride type can be written as

$$\left[\left[\text{Cs } 000, \text{Cl } \frac{1}{2}\frac{1}{2}\frac{1}{2}\right]\right].$$

The cell contains one molecule. Substituting the values of the basis into (2.13) and applying the Euler formula (1.2), we obtain a relation for the square of the structural amplitude

$$|F|^2 = [f_{Cs} \exp(-M_{Cs}) + f_{Cl} \exp(-M_{Cl}) \cos \pi(h + k + l)]^2. \quad (2.16)$$

One can see from the obtained formula that, for interferences with an even sum of the indices, $(h + k + l = 2n)$, the structural factor is equal to

$$|F|^2 = [f_{Cs} \exp(-M_{Cs}) + f_{Cl} \exp(-M_{Cl})]^2. \qquad (2.17)$$

If the sum of the indices is odd, $(h + k + l = 2n + 1)$, then

$$|F|^2 = [f_{Cs} \exp(-M_{Cs}) - f_{Cl} \exp(-M_{Cl})]^2. \qquad (2.18)$$

By measuring the intensity of interferences with even and odd sums of the Miller indices in the experiment, one can construct their dependence on the quantity $\sin^2\theta/\lambda^2$. Then, adding and subtracting the corresponding y-coordinates, one can calculate the products $f_{Cs}\exp(-M_{Cs})$ and $f_{Cl}\exp(-M_{Cl})$ separately. Since the atomic factors and their angular dependence are known, one can, using formula (2.11), easily calculate the mean-square atomic displacements separately for the atoms of each kind. This method was repeatedly applied for crystals with body-centred crystal lattice [8].

The author of this book has developed a method for separate determination of the atomic displacements in double compounds based on measurements of intensity of one and the same X-ray reflection for two temperatures. The method was applied, in particular, to the intermetallic compound Ni_3Al. The phase Ni_3Al has the ordered structure $L1_2$. The aluminum atoms are placed at the vertices of the cubic cell and form a sublattice A, the nickel atoms are at the centres of the faces and form a sublattice B. The phase has remarkable properties, in particular, an anomalous temperature dependence of the strength and is an important structural component of many heat-resistant alloys. It is known that,in accordance with the extinction rules, non-ordered face-centred structure gives reflects only if all indices are even or all ones are odd. In the ordered state, superstructural, "forbidden" reflects additionally appear. The usual reflexes of the intermetallide are (111), (200), and (220); superstructural ones are (100), (110), (210), and (211). Suppose that the intensity of each interference is measured for two temperature — high (we used the interval 723–1123 K) and the room one. From formula (2.13) we obtain the ratio of intensities of one and the same superstructural line (the subscript s) for two temperatures

$$\sqrt{\frac{I'_s}{I_s}} = \frac{-\overline{f_B}\exp(-M'_B) + \overline{f_A}\exp(-M'_A)}{-\overline{f_B}\exp(-M_B) + \overline{f_A}\exp(-M_A)}\sqrt{\frac{L'}{L}}. \tag{2.19}$$

The primed quantities refer to the high temperature; those without primes, to room temperature. The values of the exponents are given by formulae (2.11) for each atom. Since $\theta' \cong \theta$, we have $L' \cong L$. For the usual lines (the subscript u) we have

$$\sqrt{\frac{I'_u}{I_u}} = \frac{3\overline{f_B}\exp(-M'_B) + \overline{f_A}\exp(-M'_A)}{3\overline{f_B}\exp(-M_B) + \overline{f_A}\exp(-M_A)}\sqrt{\frac{L'}{L}}. \tag{2.20}$$

Denote $\sqrt{I'_s/I_s} = y_s$, $\sqrt{I'_u/I_u} = y_u$, and $\sin^2\theta/\lambda^2 = x$.

Differentiating formulae (2.19) and (2.20) with respect to x, after some transformations, at the point $x = 0$, we obtain

$$\overline{\Delta u_B^2} = \frac{3}{32\pi^2 \overline{f_B(0)}} \left\{ -\left[\overline{f_B(0)} - \overline{f_A(0)}\right] \frac{dy_s}{dx}\bigg|_{x=0} \right.$$
$$\left. - \left[3\overline{f_B(0)} + \overline{f_A(0)}\right] \frac{dy_u}{dx}\bigg|_{x=0} \right\}, \qquad (2.21)$$

$$\overline{\Delta u_A^2} = \frac{3}{32\pi^2 \overline{f_A(0)}} \left\{ \left[3\overline{f_B(0)} - \overline{f_A(0)}\right] \frac{dy_s}{dx}\bigg|_{x=0} \right.$$
$$\left. - \left[3\overline{f_B(0)} + \overline{f_A(0)}\right] \frac{dy_u}{dx}\bigg|_{x=0} \right\}, \qquad (2.22)$$

where $\overline{\Delta u_B^2}$ and $\overline{\Delta u_A^2}$ are required differences of the mean-square amplitudes of atomic vibrations in the sublattices B and A, respectively. One can see from these formulae that, to determine the differences of the displacements, we should construct experimental dependence $y_s(x)$ and $y_u(x)$ and find the derivatives at the origin. The dependences obtained for different temperatures for the considered intermetallide are shown in Fig. 2.2. As one can expect from formulae (2.21) and (2.22), the experimental dependences are linear and the determination of the values of the derivatives at the origin can be made with rather high accuracy.

For any interference, the mean values of the atomic factors of scattering can be found by the formulae

$$\overline{f_B} = \frac{1}{3}\sum_{i=1}^{m} f_i a_i; \qquad \overline{f_A} = \frac{1}{3}\sum_{j=1}^{n} f_j a_j, \qquad (2.23)$$

where f_i is the coefficient of the atomic scattering for the given angle of reflection of the ith element in the sublattice B; a_i is the atomic part of this element in the same sublattice; m is the number of elements occupying the places B. The quantities f_j, a_j and n have the similar meaning for the atoms of the sublattice A. The angular dependences of the factors of the atomic scattering are known [12].

A similar method for processing the measurements data of the interference intensity was applied in our study of vanadium carbide

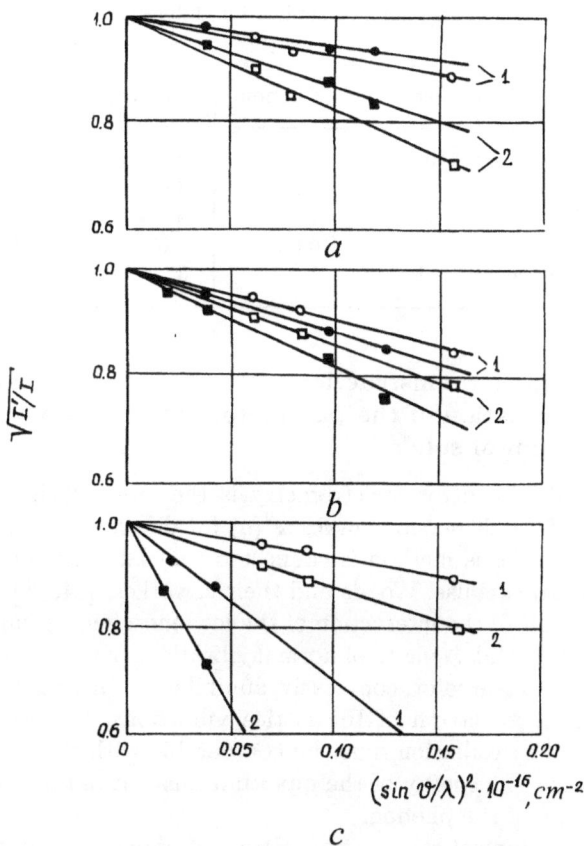

FIGURE 2.2. Experimental angular dependence of the X-ray line intensities ratio for high and room temperatures. Phase on the base of the intermetallide Ni_3Al: a — the phase $(Ni_{2.96}Fe_{0.02})Al_{1.00}$; b — $(Ni_{2.95}Fe_{0.05})(Al_{0.85}Ti_{0.15})$; c — phase isolated electrolytically from the heat-resistant alloy of mark EI-867. 1 — the measurement temperature equals 723 K; 2 — the temperature is 1023 K; o and □ — the usual reflections; • and ■ — superstructural reflections.

[13]. This allowed us to determine separately the mean-square amplitudes of vibrations of the vanadium and carbon atoms for high temperatures.

TABLE 1. Quantities typical for different quanta

Parameter	Phonons	Neutrons	X-rays
Wave length, nm	0.5	0.2	0.2
Wave vector, 10^{10} 1/m	1.2	3.0	3.0
Frequency, THz	7	10	10^6
Energy, eV	0.03	0.02	$2 \cdot 10^5$
Impulse, 10^{-24} kg·m/s	1	3	6

2.2. Methods for construction of the dispersion curves and determination of the parameters of the frequency spectrum of solids

The method of neutron spectrometry is the most efficient tool for the study of the dispersion curves $\omega^2(q)$ for different materials. The development of this method is connected to a large extent with the studies by Brockhouse, Woods and their coworkers [14, 15].

As a result of the interaction of the low-energetic, so-called heat, neutrons with solids, quanta of normal vibrations of the crystal lattice (phonons) are created or, conversely, annihilated. Naturally, to study the frequency spectrum of the lattice vibrations, the wave length of the sounding radiation must be comparable with the interatomic distances and the energy of the quantum must have the same order as the energy of the phonon.

Compare typical average parameters of phonons, heat neutrons, and X-rays. The orders of values of the corresponding characteristics are given in Table 1. Of course, these quantities are varied in some regions, and the wave lengths and the impulse values for the three types of radiation are comparable with each other (see Table 1). However, the energy of the X-ray quantum exceeds the energy of the phonon in the crystal lattice by the factor 10^7. Hence, the loss of the energy of a quantum of X-ray radiation caused by its collision with a phonon must be insignificant.

The necessary condition of the interaction is satisfied by the heat neutrons. Obviously, the collision neutron – phonon must essentially change the state of the neutron and this change can be fixed in experiments. (Neutrons are scattered by nuclei in contrast to the X-rays which interact with electron shells of the atoms.)

There are two types of scattering the neutrons on the phonons: elastic and inelastic. After an elastic collision, the mechanical energy of the neutron is not changed (but the impulse does not remain constant); after an inelastic collision, the energy is changed.

The relation between the impulse of a quantum particle and the corresponding wave length is given by the de Broglie formula:

$$\lambda = \frac{h}{mv}, \qquad (2.24)$$

where h is the Planck constant; m is the particle mass; v is its velocity. From this formula, one can readily express the particle impulse via its wave vector

$$mv = \hbar k. \qquad (2.25)$$

The energy of a quantum particle is proportional to the square of the wave vector,

$$E = \frac{\hbar^2 k^2}{2m}. \qquad (2.26)$$

Figure 2.3 illustrates the interaction of slow neutrons with a crystal lattice [85]. It is convenient to consider this interaction in the reciprocal space. By $\mathbf{k_0}$ we denote the vector of the initial flow of neutrons determining the direction of its motion; by $\mathbf{k}$, the vector of elastic scattering of neutrons. The length of each of them is equal to $2\pi/\lambda$. The equality of the lengths follows from the law of conservation of energy of neutron (2.26).

One can see from Fig. 2.3a that the condition of the elastic scattering can be written as

$$\mathbf{k} - \mathbf{k_0} = \mathbf{B}, \qquad (2.27)$$

where $\mathbf{B}$ is a vector of the reciprocal lattice. The variation of the neutron impulse has the following form

$$m\mathbf{v} - m\mathbf{v_0} = \hbar\mathbf{k} - \hbar\mathbf{k_0}. \qquad (2.28)$$

For inelastic neutron scattering, the energy of the neutron as well as that of phonon is changed; hence, the frequency corresponding to the phonon must also vary. Here, the modulus of the neutron impulse $\hbar k$ may increase as well as decrease in comparison with $\hbar k_0$. The law of conservation of impulse in inelastic collision of the neutron with

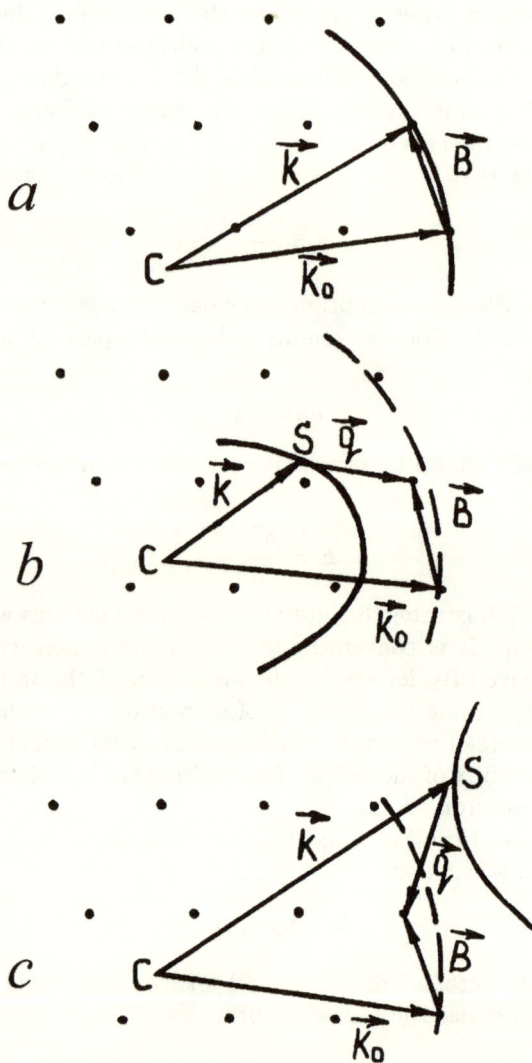

FIGURE 2.3. The schemes of interaction of the heat neutrons with the phonons (quanta of elastic vibrations of the lattice): a — an elastic collision of the neutron and phonon; b — an inelastic collision, the neutron losses the energy, a new phonon arises; c — an inelastic collision, the neutron energy grows, the phonon energy decreases.

the phonon of the lattice can be written as

$$(\hbar\mathbf{k} - \hbar\mathbf{k}_0) + \hbar\mathbf{q} = \hbar\mathbf{B}, \tag{2.29}$$

$$\mathbf{Q} + \mathbf{q} = \mathbf{B}. \tag{2.30}$$

Two cases are possible: the neutron losses the energy emitting the phonon (Fig. 2.3b); the neutron energy increases, the phonon is absorbed (Fig. 2.3c).

The law of conservation of energy can be written in the following way:

$$\frac{\hbar^2(k^2 - k_0^2)}{2m} = \pm\hbar\omega_j(\mathbf{q}), \tag{2.31}$$

where plus on the right-hand side means that the birth of the phonon occurs, and minus corresponds to its absorption. Here, $\omega_j(\mathbf{q})$ denotes the angular frequency of the oscillatory mode $(j, \mathbf{q})$; $1 \le j \le n$, where n is the number of atoms in the elementary cell; $\mathbf{q}$ is the wave vector of the appropriate oscillatory mode.

Varying the initial energy of the neutron depending on the vector $\mathbf{k}_0$, one can create phonons with various energies and, hence, frequencies. The peaks on the curve of the distribution of the incident neutrons with respect to the energies correspond to adsorbtion of neutrons and birth of appropriate phonons. The coordinate of the peak on the scale of energies allows one to find the frequency of the phonon and its wave vector $\mathbf{q}$. Repeating the measurements for different initial values of the neutron energy, one can obtain the dispersion dependence $\omega(\mathbf{q})$.

Figure 2.4 is a schematic diagram showing the principle of a triaxial neutron spectrometer. The neutron beam emerging from the nuclear reactor is reflected by the crystal-monochromator. As a result of selective, Bragg's reflection, an almost monochromatic initial neutron beam isolated from the total spectrum of energy is formed. Thus, this beam has certain values of the initial wave vector $\mathbf{k}_0$ and initial energy E_0. Then, the neutrons meet the monocrystal under investigation. The monocrystal is located in the main plane of the spectrometre and is oriented along the chosen crystallographic direction. The neutrons inelastically interact with the atomic nuclei; as a result, their wave vector and energy are changed. The values of these quantities become $\mathbf{k}$ and E. The neutron beam scattered by the crystal under investigation is analyzed with respect to energy with the help of a crystal-analyzer; the necessary angle of the rotation of the

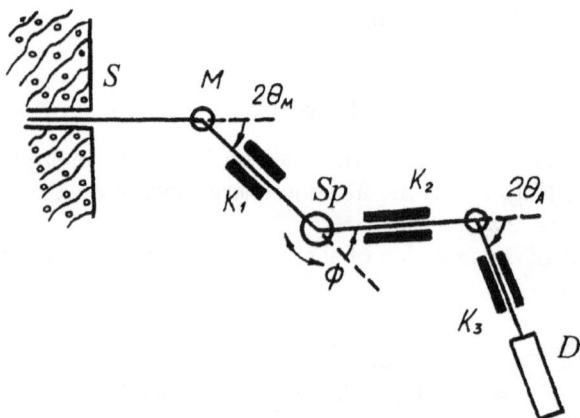

FIGURE 2.4. The scheme of a triaxial neutron spectrometer: S — the source of neutrons; M — the monochromator; Sp — the specimen; D — the detector; K_1, K_2, and K_3 — collimators.

neutron detector depends on E. Thus, while measuring, four quantities can be varied: the reflection angle from the monochromator $2\theta_M$ (change of the energy and impulse of the initial neutrons); the rotation angle of the specimen ψ (orientation of the monocrystal under investigation in the required plane and in the required direction); the angle of neutron scattering by the specimen ϕ (variation of the energy and impulse of scattered neutrons); the Woolf–Bragg angle $2\theta_A$ (measurement of the energy and impulse after scattering).

The determination of the dispersion law of the oscillatory modes can be made for fixed values of E_0 by determining the value E with the help of variation of the angle $2\theta_A$. In this method, the measurements should be repeated for various values of the wave vector of the oscillatory mode $\mathbf{q}$ of the lattice. The more efficient method, but more difficult technically is one, in which the vector of the variation of the neutron impulse preserves a constant value (so-called "method of constant $\mathbf{Q}$"). The essence of this method is described as follows. For a constant initial energy of neutrons E_0, the angles of scattering ϕ and orientation ψ of the crystal are varied step by step nonlinearly so that the vector $\mathbf{Q} = \mathbf{k} - \mathbf{k}_0$ remains constant. And, in accordance with equation (2.29), the energy $\hbar\omega$, frequency ω, and vector q are determined in the distribution of the scattered neutrons with respect to the energies. The increments of the angles ϕ and ψ are realized by a special program with the help of a control computer. First of all, the measurements are taken along the symmetric directions of the lattice $[\zeta 00]$, $[\zeta\zeta 0]$, $[\zeta\zeta\zeta]$.

The cross-section of neutron scattering by the atomic nuclei is proportional to the Bose–Einstein probability function $\dfrac{1}{\exp(\frac{h\nu}{kT}) - 1}$.
As the frequency increases, this factor rapidly decreases. This is the reason, why the experiments with the loss of the neutron energy and birth of the phonon are preferable. It is usually expedient to make the measurements under low temperatures in order to avoid the growth of inverse scattering and the variation of the peak widths due to the Debye–Waller factor.

The peculiarities of the methods plays an important role while choosing the materials for investigations. The objects must have a large capture cross-section of the heat neutrons and be available in the form of monocrystals. These requirements are met, for example, by Al, Pb, Na, Nb, Si, Ge.

3. MEAN-SQUARE AMPLITUDES OF ATOMIC VIBRATIONS IN METALLIC SYSTEMS

3.1. Pure metals and solid solutions

The dynamic atomic displacements were first treated as a measure of the forces of interatomic bonds in a crystal lattice in the 1950s by G.V. Kurdyumov and his coworkers [9, 10, 16]. The authors radiographically determined the difference of the squares of the atomic displacements in alloys for room temperature and for liquid nitrogen temperature. Most of the studies presented by these authors investigate iron and iron based alloys. The authors showed that the bond forces in the crystal lattice of ferrite are affected by alloying, plastic deformation, and heat treatment. From the author's data, all alloying elements, except vanadium, increases the bond forces in ferrite. For the temperature 296 K, under the influence of the alloying elements, the amplitudes of the atomic vibrations decrease by 18–30 %. These data refer to the atomic concentrations of the second element in iron of the order 1–2 %. For example, the addition of 8 % of chromium to alpha iron lowers the mean-square amplitude from 11.5 to 8.9 pm. The alloy containing 16 % of chromium does not differ in this characteristic from the alloy with 8 % of chromium.

We have studied solid solutions of different elements in nickel and in iron. These two metals remain the base of the materials of modern engineering. Therefore, the study of the atomic displacements in one-phase alloys has great practical value. The processes in the solid solutions should proceed simpler than in multiphase alloys. In contrast with previous studies, we made measurements of the mean-square amplitude in the temperature range 673–1023 K. The region

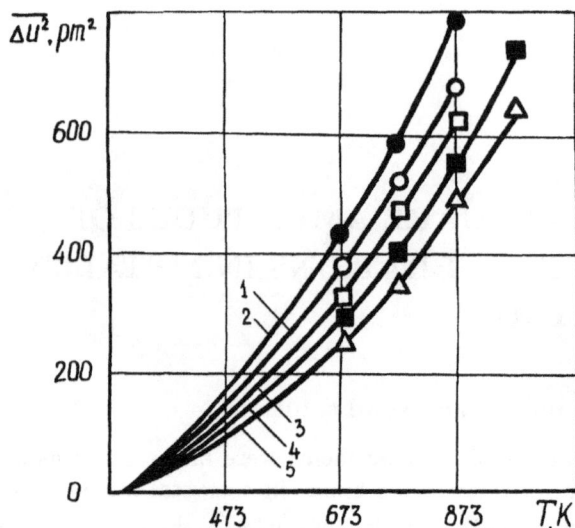

FIGURE 3.1. Influence of the temperature on the mean-square amplitude of the vibrations of atoms in solid solutions of vanadium and chromium in iron:
1 – α-iron; 2 – Fe + 4.4V; 3 – Fe + 1.4Cr; 4 – Fe + 3.9Cr; 5 – Fe + 8.6Cr.

of high temperatures plays an important role for some characteristics such as the high-temperature strength, resistance to rupture, the speed of self-diffusion.

Figure 3.1 shows the influence of the chromium and vanadium concentrations in alpha iron on the mean-square atomic displacements in the range of working temperatures 673–973 K. Vanadium increases the amplitude of the atomic vibrations and this is confirmed by the data of the works cited above. Additions of chromium in the range 1.4–8.6 of mass per cents decrease the value of $\overline{\Delta u^2}$ by 40–160 pm^2. One can see nonlinearity of the dependence $\overline{\Delta u^2}(T)$ that, obviously, is caused by the anharmonicity phenomena. Fig. 3.2 shows the influence of molybdenum. The amplitudes of the vibrations are reduced to a greater extent than in the case of addition of chromium to iron. The influence of tungsten is even greater (Fig. 3.3), and the influence of the addition of 7 % W causes the mean-square amplitudes diminish threefold.

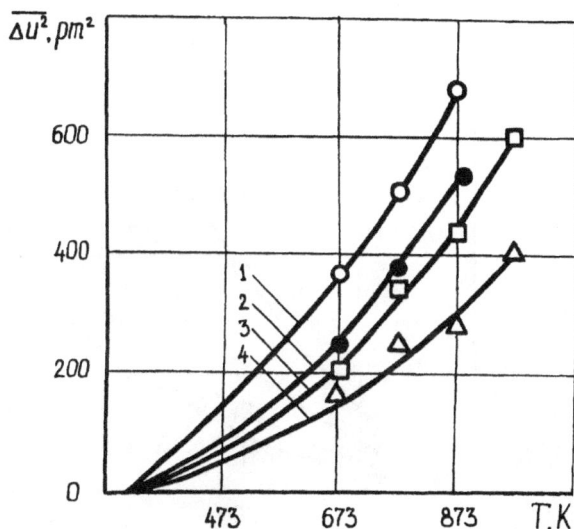

FIGURE 3.2. The same as in Fig. 3.1 for solid solutions of molybdenum in iron:
1 – α-iron; 2 – Fe + 0.7Mo; 3 – Fe + 1.2Mo; 4 – Fe + 3.4Mo.

We emphasize that, in the alloys Fe–Mo, Fe–W, the dependence $\overline{u^2}(T)$ is linear to the temperature 973 K, and the slope of the line determined by the force constant is less than in pure iron.

Cobalt also reduces the values of $\overline{\Delta u^2}(T)$, but its influence is similar to that of chromium (Fig. 3.4).

It follows from Fig. 3.5 that addition of other atoms to the crystal lattice changes the vibration amplitude measured at the temperature 873 K. The dependence of the square of the amplitude on the content of the second element is either linear (for vanadium, chromium, and silicon) or logarithmic (for cobalt, molybdenum, and tungsten). Note that we are dealing with atomic concentrations up to 6–7 %.

The composition of the investigated nickel based alloys is given in Table 1.

In addition to the two-component alloys of nickel with chromium, aluminium, and tungsten, the intermetallide Ni_3Al and four heat-resistant alloys widely used in engineering were also investigated. Table 1 shows that the least alloyed — EI437B — contains chromium

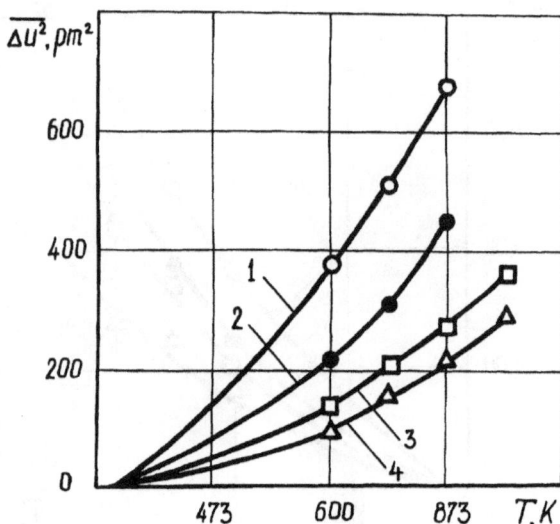

FIGURE 3.3. The same as in Fig. 3.1 for solid solutions of tungsten in iron: 1 – α-iron; 2 – Fe + 0.5W; 3 – Fe + 1.7W; 4 – Fe + 2.7W.

TABLE 1. Chemical composition of investigated nickel based alloys, mass %% (C < 0.06; Mn < 0.25; Si < 0.30)

Material	Cr	Al	Ti	W	Mo	Co	B	Nb
Ni–Cr	8.5	—	—	—	—	—	—	—
Ni–Al	—	4.83	—	—	—	—	—	—
Ni–W	—	—	—	24.8	—	—	—	—
Ni$_3$Al	—	12.62	—	—	—	—	—	—
EI437B	20.1	0.70	2.52	—	—	—	0.006	—
EI698	14.0	1.65	2.70	—	2.99	—	0.003	2.04
EI867	9.5	4.47	—	5.3	9.82	5.12	0.020	—
EI199	19.8	2.14	1.42	9.1	4.54	—	0.006	—

as well as titanium and aluminium; also in this alloy, the Ni$_3$(Al,Ti) phase is formed. The scheme of alloying the three last alloys includes

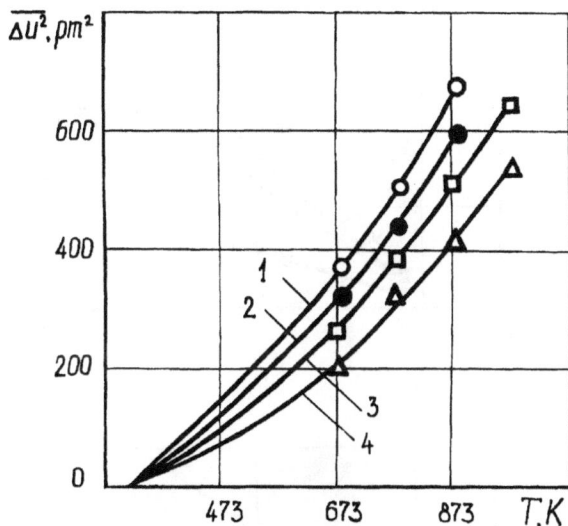

FIGURE 3.4. The same as in Fig. 3.1 for solid solutions of cobalt in iron: 1 – α-iron; 2 – Fe + 1.0Co; 3 – Fe + 3.0Co; 4 – Fe + 7.8Co.

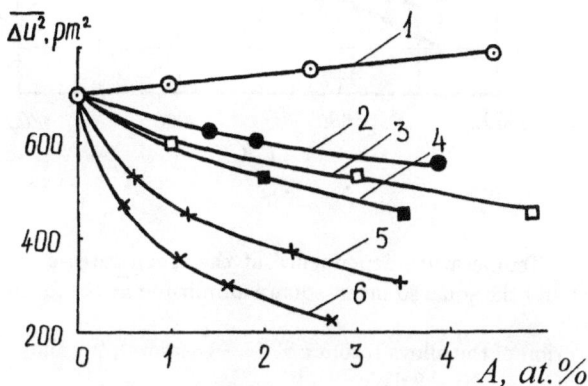

FIGURE 3.5. Influence of the atomic concentration of the second element in the iron alloys on the mean-square amplitudes at the temperature 873 K: 1 – V; 2 – Cr; 3 – Co; 4 – Mn; 5 – Mo; 6 – W.

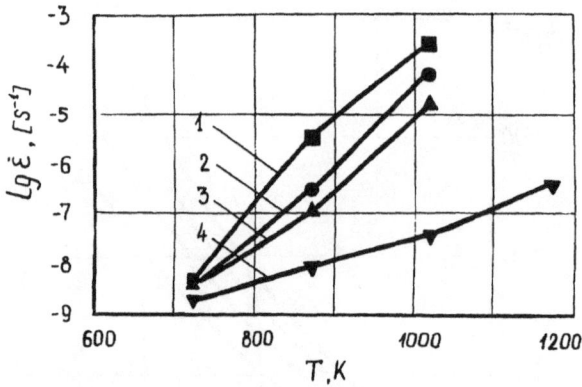

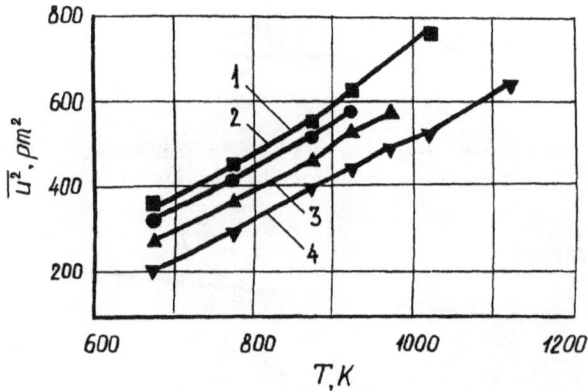

FIGURE 3.6. Temperature dependence of the creep rates $\dot{\varepsilon}$ at the stress 80 MPa (a) and the squared mean-square amplitude $\overline{u^2}$ (b) for nickel based alloys:
the composition of the alloys (atomic %%) – 1 – nickel; 2 – alloy Ni + 9.9Al; 3 – Ni + 9.5Cr; 4 – Ni + 9.6W.

complex additions of other elements in their composition. The experiments show that the high-temperature strength increases, the creep rate decreases, and the values of the working temperatures grow.

In Fig. 3.6, one can see the influence of temperature and the composition of nickel based solid solutions on the mean-square displacements of the atoms. Pure nickel has the greatest values of the mean-square displacements for high temperatures. The addition of other elements results in decrease of the square of the mean-square amplitude $\overline{\Delta u^2}$ for high temperatures. The effect of the alloying increases in the series chromium–aluminium–tungsten. The most efficient influence on the increase of the bonds between atoms in solid solution is caused by the addition of tungsten. In this case, after adding 10 atomic % W, the mean atomic mass increases in 1.2 times, but the values of the differences of the squared displacements decrease, depending on the temperature, in 1.6–1.8 times. Hence, the tungsten atoms in the crystal lattice of nickel not only lower the amplitudes of the vibrations, by virtue of increasing the mean atomic mass, but also reinforce the interatomic bonds.

The comparison of the temperature dependence of the differences of squared displacements and rates $\dot{\varepsilon}$ of the steady-state creep of nickel and its solid solutions (Fig. 3.6) clearly shows the connection between the two quantities. Correlate the amplitude of atomic displacements with one of the main characteristics of the high-temperature strength — the rate of the stationary creep. Naturally, the less is the creep rate, the greater is the high-temperature strength. One can see in Fig. 3.6 that, in the temperature range 850–1050 K, i.e., for the temperatures where the mechanism of the creep is controlled by diffusion in the solid solutions, the correlation between the creep rate and amplitudes of the heat atomic vibrations (and, hence, diffusion characteristics of the material) can be observed. For a relatively low testing temperature 673 K, the mechanism of deformation is determined not only by the diffusion, but to a large extent by slip of dislocations dominates. It is likely to be the reason why, for this temperature, the creep rates of nickel and its solid solutions are practically equal each other.

The rate of steady-state creep can be expressed [17] as

$$\dot{\varepsilon} = \dot{\varepsilon}_0 \exp\left(-\frac{U_0 - \gamma\sigma}{kT}\right), \qquad (3.1)$$

where U_0 is the energy of the self-diffusion activation; γ is the activation volume; $\dot{\varepsilon}_0$ is the pre-exponential factor; σ is the applied stress; k and T have their usual meanings.

The decrease of the vibration amplitude under the action of dissolving the foreign atoms in nickel increases the value of the energy of the diffusion activation U_0 in the crystal lattice that, in turn, must reduce $\dot{\varepsilon}$. The most noticeable is the reduction of the creep rate (by three–four orders of the magnitude for temperatures 850–1050 K) under the action of the tungsten additions.

In the approximation of the pairwise interaction of the atoms, the force constant Φ'' can be determined starting from the experimental dependence of the squared atomic displacements on the temperature. Indeed, expanding the potential energy of the lattice into series (1.7), we have (the first term on the right-hand side disappears while transferring the origin, the second one is equal to zero, the fourth one is neglected in the harmonic approximation):

$$\Phi = \frac{1}{2}\Phi''\overline{u_x^2}. \tag{3.2}$$

The squared displacement in any direction is $\overline{u^2} = 3\overline{u_x^2}$. In the region of high temperatures, the potential energy of the node per one degree of freedom is equal to $\frac{1}{2}kT$. Differentiating (3.2) with respect to temperature and denoting the force constant $\Phi'' = \alpha$, we obtain

$$\alpha = \frac{3k}{d\overline{u^2}/dT}. \tag{3.3}$$

Thus, in order to estimate the value of the force constant for the above approximations, the slope of the curve of dependence of the square of mean-square displacement on the temperature should be determined. The force constant is inversely proportional to the slope of the curve.

Table 2 shows the results of calculation of the force constant α for a series of metals with cubic face-centred or cubic body-centred crystal lattice. The experimental data obtained in a series of investigations are used [18].

Table 2 also shows that this constant has the greatest values for chromium and tungsten. The least value is obtained for sodium and lead.

It has been described in the first section that the value $\overline{u^2}$ is inversely proportional to the atom mass m (formula (1.61)). The vibrations of atoms of different mass can be compared with each

TABLE 2. Coefficients of the temperature dependence of the square of the mean-square amplitude of some metals and the force constants of the pair atomic interactions

Metal	Z	$\dfrac{\overline{du^2}}{dT}, \dfrac{\text{pm}^2}{\text{K}}$	$\alpha, \dfrac{\text{N}}{\text{m}}$	$\eta, \dfrac{10^{-26}\ \text{kg}\cdot\text{pm}^2}{\text{K}}$
Na	11	9.72	4.3	37.3
Al	13	1.46	28.4	6.5
Cr	24	0.32	127.7	2.8
Fe	26	0.50	82.3	4.7
Ni	28	0.50	82.8	4.9
Cu	29	0.74	55.7	7.9
Ag	47	1.20	34.4	21.7
W	74	0.26	161.1	7.9
Pb	82	47.87	8.8	161.2
Au	99	8.20	50.5	27.0

other, but it is expedient to exclude its effect by multiplying the values $\dfrac{\overline{du^2}}{dT}$ by the mass of one atom.

Thus, we introduce the reduced displacement — the parameter η which characterizes the forces of interatomic bonds

$$\eta = m\frac{\overline{du^2}}{dT}. \tag{3.4}$$

As the bonds between the atoms weaken, this parameter increases.

The values of η are presented in the last column of Table 2. Ordering the metal by increasing of the value of the reduced displacement, we obtain the following series: Cr, Fe, Ni, Al, W, Cu, ... Table 2 shows the possibilities of chromium as a material with strong interatomic bonds; iron, nickel, aluminium, and tungsten also have favourable values of the parameter η.

The mean-square amplitudes of the atomic vibrations are rather sensitive to the composition of the triple alloys. The amplitudes for alloys of the system Ag–Cd–In are presented [19]. (The concentrations are given in atomic per cents.) The three elements are neighbours in the periodic system. The difference of atomic masses are rather small. Silver is univalent, cadmium is divalent, and the typical valency of indium is equal to three.

Ag – Cd – In	$\sqrt{\overline{u^2}}$, pm	Δz
100 – 0 – 0	14.3	1.00
90 – 5 – 5	17.4	1.50
80 – 10 – 10	20.1	1.50
80 – 15 – 5	20.5	1.25
85 – 10 – 5	20.6	1.33
75 – 20 – 5	23.3	1.20

One can see that, as the atomic concentrations of additions foreign for silver increase, the mean-square amplitude also increases, obviously, because the interatomic bonds weaken. Also the correlation is observed between the quantity $\sqrt{\overline{u^2}}$ and the cadmium concentration in the alloys. The authors associate the influence of the alloys composition upon the vibrations amplitude with the difference in the valency of the solvent — silver and that of the solved elements. To take this influence into account, the authors suggested the introduction of the parameter Δz that is calculated in the following way:

$$\Delta z = \frac{a_{Cd} \times 1 + a_{In} \times 2}{a_{Cd} + a_{In}}, \qquad (3.5)$$

where a is the atomic part of the element. One and two in the numerator correspond to the difference in the valency of the solved element and that of the solvent. However, the correlation between the heat displacements and the introduced quantity is inobservable. There is no correlation with the electron concentration.

Similar results are obtained for the alloys of the system Ag–Cd–Zn [20]:

Ag – Cd – Zn	$\sqrt{\overline{u^2}}$, pm
100 – 0 – 0	14.3
90 – 5 – 5	15.6
90 – 2 – 8	16.0
90 – 5 – 15	16.4
70 – 10 – 20	17.0
80 – 16 – 4	17.3
70 – 29 – 1	18.1

TABLE 3. Parameters of the alloys of aluminium with nickel and cobalt

Alloy	Content of the second element, at. %					
	54		50		46	
	$\overline{du^2}/dT$, pm^2/K	α, N/m	$\overline{du^2}/dT$, pm^2/K	α, N/m	$\overline{du^2}/dT$, pm^2/K	α, N/m
Al–Ni	0.71	58.1	0.65	63.7	0.54	77.0
Al–Co	1.13	36.8	0.98	42.5	0.78	53.4

In [21], the temperature dependence of the mean-square amplitudes for the alloys Al–Co and Al–Ni are measured. These intermetallides have an ordered structure of type CsCl. The study was made in the interval of atomic concentrations 46–54 %. It was shown that alloying aluminium with nickel lowers the amplitudes in greater degree than alloying by cobalt. Table 3 presents the obtained characteristics of the alloys.

3.2. Anisotropy of atomic vibrations

The properties of the crystal lattice are unequal in different crystallographic directions (the anisotropy phenomenon). This fact reflects the quality of the interatomic bonds of metallic atoms such as its directivity. The experimentally measured mean-square amplitudes of the atomic vibrations are different in different directions.

The anisotropy of the atomic vibrations in metals was studied for elements with compact hexagonal crystal lattice. The ratio of the axes c/a in a hexagonal lattice is usually in the range 1.45–1.59. In the published papers, serious attention was given to the little-studied group of the rare-earth metals.

In Table 4, some experimental results are presented. The highest degree of anisotropy, essentially exceeding the measurement errors of the mean-square displacements, is proper to titanium, zinc, cadmium, and europium. For zinc, the displacements perpendicular to the plane of the base of the hexagonal lattice essentially exceed those in the plane of the base. Clearly, the interatomic forces are different in these two directions — they are less along the axis of the sixth order symmetry (c-axis). In the crystal lattice α–Ti (low-temperature modification), the anisotropy of displacements is present in all the

TABLE 4. Anisotropy of the heat atomic vibrations in metals with hexagonal compact crystal lattice. Room temperature

Metal	Z	$\sqrt{u^2}$, pm		Source
		$\| c$	$\perp c$	
Mg	12	15.6	15.6	[22]
Sc	21	9.6	9.5	[23]
Ti	22	10.7	12.3	[24]
Zn	30	19.9	13.0	[22]
Y	39	10.1	10.2	[25]
Cd	48	20.2	11.7	[87]
Gd	64	10.6	10.1	[25]
Tb	65	9.5	9.2	[23]
Dy	66	10.6	10.1	[25]
Er	68	11.8	14.5	[26]
Lu	71	11.1	10.4	[25]

studied temperature interval from 80 to 1000 K, but the degree of anisotropy decreases as the temperature increases [24]. While investigating the monocrystals of cadmium in the temperature interval 78–300 K [27], it was found out that amplitudes of the atomic vibrations along the c-axis are essentially greater than those along the a-axis lying in the basis plane. The temperature coefficient is also greater in the first case.

In terbium, dysprosium, and lutecium, the results are similar — $\sqrt{u_\perp^2} < \sqrt{u_\|^2}$ — but the difference is small. In magnesium and scandium, anisotropy is not detected.

Thus, the type of the crystal lattice does not itself determine the relation between the vibration amplitudes in the mutually perpendicular directions. It seems that the nature of the metal plays a certain role.

In our work, we consider iron and the studied solid solutions on its base contained about one atomic per cent of tungsten or molybdenum. The amplitude of the atomic vibrations in the temperature interval 293–953 K was determined by the method of X-ray analysis in directions perpendicular to the following crystallographic planes: (100), (110), and (211).

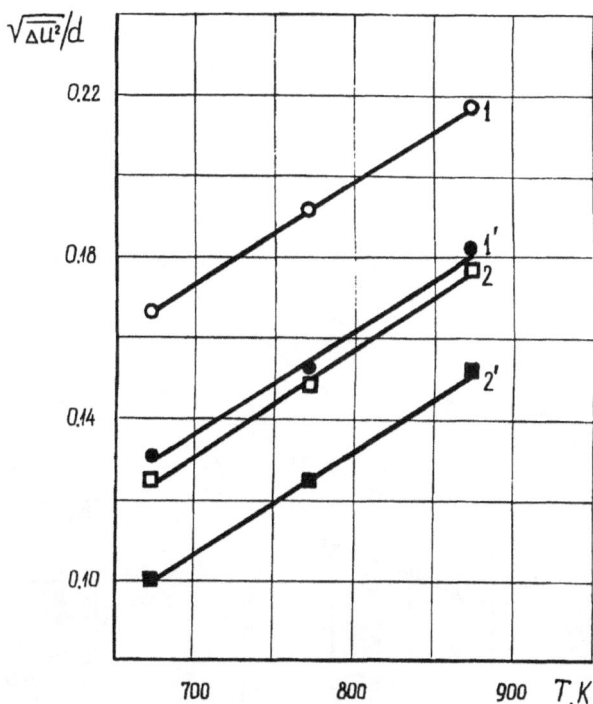

FIGURE 3.7. Plot of the relative atomic displacements versus temperature: 1, 2 — α-Fe; 1′, 2′ — α-Fe + 0.34%W; 1, 1′ — the direction of the atomic vibrations is <211>; 2, 2′ — the direction of the atomic vibrations is <100>.

The results obtained imply that the heat atomic vibrations in iron and in the studied solid solutions differ by clearly defined anisotropy. In Figs. 3.7 and 3.8, the experimental data are presented. In the graphs, along the y-axis we plot the relative displacements, i.e., the ratio of $\sqrt{\overline{u^2}}$ to the value of the appropriate measured interplane distance d. One can see that the relative atomic displacements linearly depend on the temperature in the considered interval and achieve quite large values: $(0.1–0.2)d$. The graphs imply that very small additions (only 3–6 atoms of W or Mo per 1000 atoms of iron) essentially lower the amplitude of the relative displacements of vibrating atoms. And, tungsten affects more strongly than molybdenum. The least relative displacements of atoms are observed for vibrations along

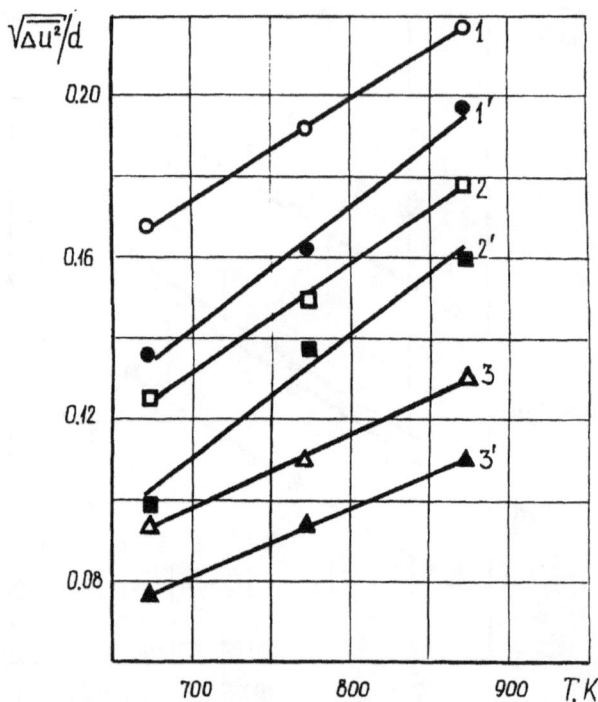

FIGURE 3.8. Plot of the relative atomic displacements versus temperature: 1, 2, 3 — α-Fe; 1′, 2′, 3′ — α-Fe+0.67%Mo; 1, 1′ — the direction is <211>; 2, 2′ — the direction is <100>; 3, 3′ — the direction is <110>.

⟨110⟩ and the greatest, along ⟨211⟩. The alloying elements have little effect on the degree of anisotropy of the displacements; the increase in temperature within the studied limits lowers the influence of molybdenum on the vibrations along the directions ⟨211⟩ and ⟨110⟩.

In Table 5, some calculated parameters of packing are given for three crystallographic planes of the lattice of alpha iron: the interplane distances d, the areas of an elementary plane cell S; the numbers of atoms in a plane cell n; the surface densities of atoms ρ.

Comparing the graphs in Figs. 3.7 and 3.8 with data in Table 5, we come to the conclusion that the least value of the relative mean-square amplitude of heat atomic vibrations is observed in perpendicular to the most closely packed plane (110). Conversely, the maximal

TABLE 5. Density of the packing of the atomic planes in the crystal lattice of alpha iron

(hkl)	d, 10^{-10} m	S, 10^{-19} m^2	n	ρ, 10^{19} m^{-2}
(211)	1.18	2.00	2	1.00
(100)	1.44	0.82	1	1.22
(110)	2.04	1.16	2	1.73

relative amplitude is typical for the least closely packed atomic plane (211).

Now, we consider why small additions of W and Mo to iron essentially lower the mean-square amplitude. The ratio of atomic masses of tungsten and iron is equal to 3.29; that of molybdenum and iron is equal to 1.71. However, the observed phenomenon cannot be explained by the increase of the atomic mass. If one takes into account the atomic part of the second element in the materials investigated, then one sees that the average atomic mass changes after dissolution quite insignificantly, approximately by 0.8 %. At the same time, the obtained data imply that the vibration amplitudes decrease by 20 % and more. Obviously, the situation is different. There are good grounds for believing that the addition of a small quantity of atoms of molybdenum or tungsten into iron results in a noticeable change of the oscillatory spectrum of the alloy; namely, in the growth of the part of high frequencies in the spectrum. Massive atoms, replacing the atoms of iron in the nodes of the crystal lattice of the solvent, divide the chains of consistent vibrations of the lattice into several parts and shorter atomic chains fixed at some points are obtained. The frequency of their vibrations must be higher and the amplitude of the vibrations, accordingly, must be lower than in pure iron. We have considered this problem in Subsection 1.6 of Section 1. From formula (1.60) deduced in this subsection, it is clear that the mean-square amplitude is inversely proportional to the frequency of the oscillatory mode. The emergence of the local oscillatory modes in the spectrum must essentially change the value of the mean-square amplitudes of the atomic vibrations. The second essential reason for the considered phenomenon can be the change of the force constants in the vicinity of the foreign atoms.

3.3. Intermetallic compounds

The ordered phases denoted by the formula B_3A have the superstructure $L1_1$. This means that, in the elementary cubic face-centred cell of the intermetallide, the vertices of the cube are occupied by atoms of the type A and the centres of the faces, by atoms of the type B. The chemical compound of nickel with aluminium Ni_3Al belongs to such phases. The solid solutions of a number of elements in this intermetallide are formed in the process of the so-called aging of wide-spread heat-resistant alloys on the base of nickel. The disperse particles of these phases precipitated from the solid solution and coherently connected with the matrix provide the strength properties of the alloys in the region of high temperatures. The stability of the structure and the high-temperature properties of the alloys for an applied loading depend on the properties of the disperse phase.

The discussed phase possesses an anomalous temperature dependence of the strength: its strength properties increase as the temperature rises. The nature of the deformation of phases of the type B_3A for high-temperature creep is studied insufficiently. There is good reason to believe that, in these conditions, the slip of the deforming dislocations is controlled by the diffusion processes. Indeed, with ordering, the energy of the diffusion activation and the creep resistance increase. A study of the nature of the high-temperature deformation resistance of the alloys is of interest for measurements of the amplitude of the atomic vibrations.

We have measured the mean-square amplitudes of the atomic vibrations in solid solutions on the base of Ni_3Al. As the materials for the investigation, the polycrystalline alloy Ni_3Al and strengthening phases representing solid solutions of different elements in an intermetallide of the type B_3A were used; these phases were extracted electrolytically from aged heat-resistant alloys of four marks (the composition of the alloys is given in Table 3, the four last lines, respectively).

In the second section, we described the methods developed by us for determination of the values of mean-square amplitudes for each of the two sublattices of the intermetallide separately (formulae (2.21) and (2.22) and Fig. 2.2).

In order to determine the chemical composition of the investigated phases, the results of the chemical analysis of studied powder

TABLE 6. Mean-square atomic displacements in phases of the type B_3A

Phase	$\overline{\Delta u_B^2}$, pm^2			$\overline{\Delta u_A^2}$, pm^2		
	723	873	1023 K	723	873	1023 K
$Ni_{2.96}Fe_{0.02}Al_{1.00}$	330	470	660	380	580	810
$Ni_{2.95}Fe_{0.05}(Al_{0.85}Ti_{0.15})$	380	440	600	290	370	460
$Ni_{2.83}Fe_{0.04}Cr_{0.07}Mo_{0.06}$ $(Al_{0.51}Ti_{0.31}Nb_{0.07}Mo_{0.08}Cr_{0.03})$	250	380	550	180	230	330
$Ni_{2.83}Fe_{0.04}Cr_{0.08}W_{0.03}Mo_{0.02}$ $(Al_{0.43}Ti_{0.28}W_{0.08}Mo_{0.03}Cr_{0.10})$	230	340	450	30	60	80
$Ni_{2.76}Co_{0.09}Fe_{0.01}Cr_{0.06}Mo_{0.02}$ $(Al_{0.79}W_{0.08}Mo_{0.11}Cr_{0.03})$	260	340	580	20	40	50

objects were used; with the same purpose, the results of the radiod-ifractometric measurements of the ratio of intensities of two orders of reflection of the phases (100) and (200) were analyzed. The final formulae were established by the coincidence of the experimental and calculated values of this ratio.

The results of determination of the composition of the phases and calculation of the mean-square atomic displacements are presented in Table 6. In the complexly alloyed phases, the atoms of tungsten and molybdenum are dissolved, replacing mainly the atoms of aluminium, but also partially taking the places of nickel. Chromium is distributed between the two sublattices. The table shows that, in the first phase practically unalloyed, the atomic displacements of lighter atoms of aluminium (nodes A) appreciably exceed the displacements of the atoms of nickel (nodes B) for all considered temperatures. The replacement of approximately one sixth of the aluminium atoms by the titanium atoms decreases the amplitudes of the heat vibrations in the aluminium sublattice so that we now have $\overline{\Delta u_A^2} < \overline{\Delta u_B^2}$. As the temperature rises, the effect of decrease of the displacements under the influence of the dissolution of titanium makes itself evident in higher degree. The effect of adding molybdenium and tungsten to the phase is even greater. One can see in Table 6, for instance, that, at the temperature 1023 K, both these elements lower the values $\overline{\Delta u_A^2}$ in 10-16 times. A less essential decrease of the amplitude of the vibrations is observed in the sublattice B.

The decrease of $\overline{\Delta u_A^2}$ is not proportional to the growth of the mean atomic mass, but essentially exceeds it. Actually, the effective mass of atoms A in investigated strengthening phases increases as compared with the mass of the aluminium atoms only by 64–78%; this of the B atoms is almost the same as that of nickel. This fact obviously shows the growth of the forces of interatomic bonds in the crystal lattice of the alloyed intermetallide. Comparing the values of the mean-square atomic displacements in the sublattice A for phases of different compositions, one can conclude that they are determined, mainly, by the sum of the atomic parts of tungsten and molybdenium. In Table 6, the phases are ordered by increasing this sum and a decrease of the quantity $\overline{\Delta u_A^2}$ in the same order is observed.

The decrease of the amplitude of the heat vibrations of atoms must increase the energy of the diffusion activation in the crystal lattice and decrease the creep rate if it actually is limited by the diffusion. To check this assumption, we have compared the mean-square atomic displacements in the extracted strengthening phases of the alloys with the rate of the steady-state creep $\dot{\varepsilon}$ of the same alloys. Indeed, the results obtained shows the correlation between the two considered quantities. The alloy with $\overline{\Delta u_A^2} = 330$ pm^2 at 1023 K has the least high-temperature strength among the three last alloys in Tables 1 and 6. The alloy with this quantity equal to 50 pm^2 has the greatest high-temperature strength.

The question arises as to whether the diffusive mobility of elements in the crystal lattice of an intermetallide can essentially decrease if the mean-square displacements decrease steeply only for the aluminium sublattice? Taking into account that the energetically justified mechanism of the elementary diffusive act in ordered alloys consists in a certain cycle of the vacancy shift and that the nodes of both the lattices must be included in this cycle, one can answer this question positively.

The following formula for the creep rate of heat-resistant alloys strengthened by disperse particles of the phase B_3A is justified by the authors [28]:

$$\dot{\varepsilon} = f(\bar{r}) N \exp\left(-\frac{U - \varphi(\bar{r})\,\sigma}{kT}\right), \qquad (3.6)$$

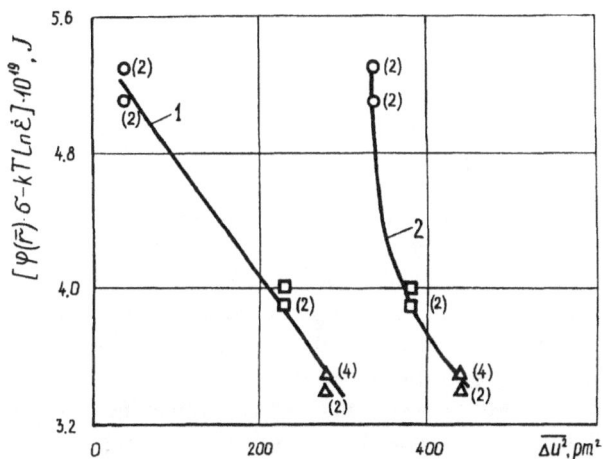

FIGURE 3.9. Relationship between the quantity characterizing the creep resistance of alloys (the first term on the right-hand side of equation (3.9)) and the mean-square amplitude of the atomic oscillations in the strengthening phase. The amplitudes were measured at the temperature 873 K: 1 — oscillations of the atoms in the sublattice A; 2 — oscillations of the atoms in the sublattice B.

where N is the density of dislocations; U is the energy of the process activation coinciding in magnitude with the energy of the diffusion activation U_D in the strengthening phase; σ is the effective applied stress; $f(\bar{r})$ and $\varphi(\bar{r})$ are functions of the size of the particles of the strengthening phase. In paper [29], an equation is proposed for the energy of the vacancy formation E_v. This energy is closely connected with the mean-square amplitude of the atomic vibrations:

$$E_v = \frac{Kma^2}{\overline{u^2}}, \qquad (3.7)$$

where K is a coefficient of proportionality; m is the mass of the atom; a is the lattice period. It seems natural that the greater is the amplitude of the atomic vibrations, the greater is the probability that some atoms overcome the potential barrier and the smaller is E_v. On the other hand, for most metals with a cubic lattice, we have

$E_v = (0.50\text{--}0.63)U_D$ [30]. Hence, we can write

$$U_D \sim \frac{1}{\overline{u^2}}. \tag{3.8}$$

Taking the logarithm of equation (3.6), we obtain

$$U = \left[\varphi(\bar{r})\,\sigma - kT\ln\dot{\varepsilon}\right] + kT\ln\left[f(\bar{r})\,N\right]. \tag{3.9}$$

The second of the two terms of sum (3.9) depends on the applied stress weakly (only logarithmically via N). If our assumptions $[U \cong U_D$ and (3.8)] are valid, then, the less are the mean-square displacements in the strengthening phase (greater U), the greater should be the first term of equation (3.9) included in the brackets.

In Fig. 3.9, the graph is based on the results of investigation of three heat-resistant alloys. The creep resistance increases in the positive direction of the y-axis. Every point on the graph corresponds to one applied stress; the numbers in the parentheses refer to the number of quantities producing coinciding points. The graph shows that, as the mean-square amplitude of atomic displacements decreases, a decrease in the creep rate is observed. This affects both the sublattices of strengthening phase.

4. AMPLITUDES OF ATOMIC VIBRATIONS IN SEMICONDUCTIVE MATERIALS AND IN CHEMICAL COMPOUNDS

4.1. Interatomic bonds and vibrations of atoms in elements of IV group of the periodic system. The nature of the covalent bond

The consideration of data concerning the heat vibrations of atoms in semiconductors begins with the elements of the first subgroup of the IV group of Mendeleev's periodic table. In ascending order of atomic mass, the elements are carbon, silicon, germanium, tin and lead.

The properties of these elements are most closely connected with the configuration of the electron shell of their atoms. It is known that the valent shells are described by the formula ns^2np^2, where the principal quantum number n varies from 2 for carbon to 6 for lead; every of the two shells contains two electrons.

In the initial state, the structure of electron shells of a carbon atom can be represented by the formula $1s^22s^22p_x^12p_y^1$. The wave functions for $1s$ and $2s$ electrons are spherically symmetric, whereas the wave function of $2p$-electrons is not spherically symmetric. The product of the squared wave function by a volume element gives the probability that the electron is in this volume. The strength of the interatomic bond formed by p–p-electrons exceeds those of the s–s-bond approximately by three times [31]. While interacting with other atoms, the carbon atom comes into exited state described by the formula $1s^22s^12p_x^12p_y^12p_z^1$. In other words, the so-called tetrahedron hybridization occurs: s- and p-electrons are mixed, hybridizated and, as a result, the atom obtain four unpaired electrons which can form four directed equivalent bonds located under equal angles to each

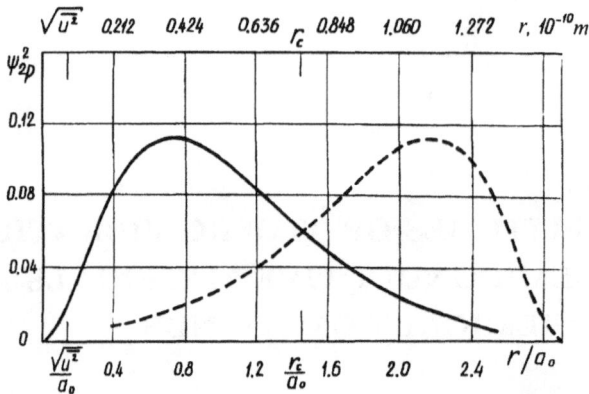

FIGURE 4.1. Dependence of the square of the wave function of electrons on the distance r for neighbouring carbon atoms in the diamond structure: r_c is the radius of the carbon atom; $a_0 = 0.53 \cdot 10^{-10}$ m is the Bohr radius.

other. The values of the wave function of the hybridizated electrons are different in different directions.

To study the explicit form of the wave functions, it is convenient to use the Slater–Siner expressions. In the case of the considered $2p$-electrons, one can write [32, 33]:

$$\psi_{2p} = \left(\frac{c^5}{32\pi}\right)^{0.5} \frac{r}{a_0} \exp\left(-\frac{cr}{2a_0}\right), \qquad (4.1)$$

where the constant $c = Z - s$; Z is the number of the element in the periodic table; s is the screening constant equal for carbon to 3.25; r is the distance from the nucleus; a_0 is the Bohr radius.

The results of the calculation of the squared wave functions by formula (4.1) for the two nearest atoms of carbon in the crystal lattice of diamond are presented in Fig. 4.1. It should be noted that, in this figure, the wave functions of free atoms are shown shifted by a distance typical for the diamond lattice. In the real interatomic bond, the joining of two wave functions of neighboring atoms changes their coordinate dependence in the interatomic region. Nevertheless, one can see that the two p-electrons, which fulfill the covalent bond, belong to both the atoms. The maximum of the squared wave function falls at the quarter of the interatomic distance. As a comparison, the graph shows, in the same scale, the value of the amplitude of the

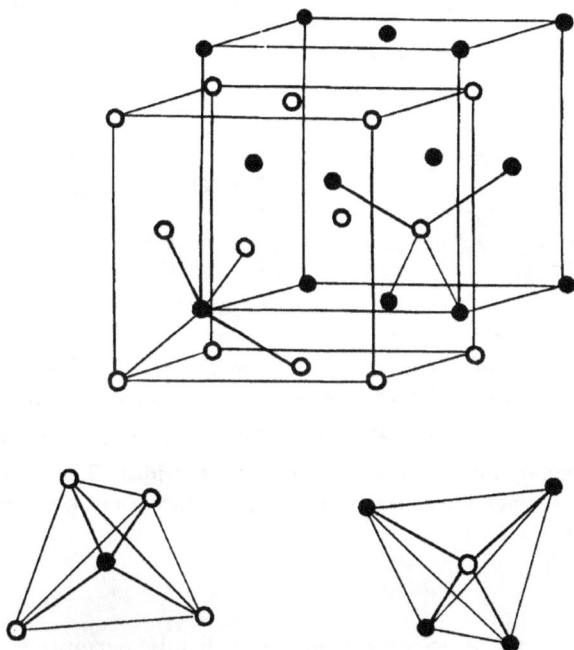

FIGURE 4.2. Location of atoms in the crystal lattice of the diamond type: A (white) and B (black) are two types of nodes of an elementary cell.

mean-square displacement of atom in diamond at the room temperature. This amplitude is relatively small and is within only 5% of the interatomic distance.

Some properties of the elements of the first subgroup of the IV group are represented in Table 4.1. All the data are related to the temperature 293 K.

The unique characteristics of diamond are stipulated by the essential forces of the covalent bond in the crystal lattice. The crystal structure of the diamond type is shown in Fig. 4.2. It can be represented as two cubic face-centred lattices, one inserted into the other, and shifted by a quarter of the solid diagonal. We give a basis of the lattice (A, B are the carbon atoms).

$$A\left(000; \frac{1}{2}\frac{1}{2}0\right), \qquad B\left(\frac{1}{4}\frac{1}{4}\frac{1}{4}; \frac{1}{4}\frac{3}{4}\frac{3}{4}\right).$$

TABLE 1. Some properties of elements of the IV group.

Element	Structure of the electron shells	K, 10^{-5} m^2/N	α, N/m	E_g, eV	θ_M, K	$\sqrt{\overline{u^2}}$, pm	$\frac{\sqrt{\overline{u^2}}}{r_{at}}$
C	$1s^2 2s^2 2p^2$	0.18	214.2	5.50	1860	7.5	0.10
Si	$1s^2 2s^2 2p^6 3s^2 3p^2$	1.02	80.4	1.12	543	9.7	0.08
Ge	$1s^2 2s^2 2p^6 3s^2 3p^6 3d^{10} 4s^2 4p^2$	1.33	56.4	0.67	290	13.8	0.11
Sn	$\ldots 5s^2 5p^2$	1.88	25.2	0.10	150	22.0	0.16
Pb	$\ldots 6s^2 6p^2$	2.37	15.0	—	88	28.3	0.19

REMARK. K is the coefficient of compressibility; $K = \frac{dV/dp}{V} = \frac{3}{c_{11}+2c_{12}}$ [34], where c_{ik} are moduli of elasticity; α is the force constant; θ_M is the X-ray Debye temperature; E_g is the width of the forbidden gap; $\sqrt{\overline{u^2}}$ is the mean-square amplitude of the atomic vibrations. The values of $\sqrt{\overline{u^2}}$ and θ_M for diamond are taken from [35]; for silicon and germanium, from [36]; for tin, from [37].

The indices of the nodes should be cyclically permuted; thus, the elementary cell contains eight equivalent atoms. However, in the semiconductive compounds that will be considered, the nodes A and B are occupied by different atoms (the structure of zinc blende). In the lattice of diamond, there are four equal covalent bonds located at the angle $109°28'$ to each other; every one is formed by a pair of electrons with opposite directed spins.

With reference to the data of Table 4.1: in the carbon atom, the screening effect of the spherical $1s$-shell is very weak; as a result, valency electrons are strongly bonded with the nucleus. The compressibility of diamond is quite small, and in a similar way the mean-square amplitude of atomic vibrations is also small. A consequence of the strong bond is also the electric properties. The electrical conductivity of diamond is practically equal to zero: the width of the forbidden gap is 5.5 eV.

The crystal structure of silicon is the same; however, here, the hybridization is made by electrons of the shells $3s^2 3p^2$. In connection with arising of internal shells, additional in comparison with the carbon atoms in diamond, the structure of silicon is more "loose". Accordingly, the coefficient of compressibility increases by almost ten times and the force constant essentially decreases. The screening effect of the internal $2p$-electron shell results in possibility of delocalization of a part of valency electrons that can be seen in the decrease of the width of the forbidden gap to 1.12 eV. Silicon is a typical semiconductor. At the room temperature, the mean-square amplitude of atomic vibrations in silicon is in 1.7 times greater than that in diamond.

The pass to germanium and then to tin illustrates the same idea, i.e., the increasing influence of the internal electron shells, which screen the charge of the nucleus, and weaken the bond of the valency electrons with the atomic nucleus. In grey tin, and especially in lead, the valency electrons are delocalized and both the elements are metals.

Now, consider the influence of the temperature on the amplitude of the atomic vibrations. In Fig. 4.3, the dependences of the value $\overline{u^2}$ on the temperature are presented for silicon and germanium. Curves 1 and 2 are calculated on the basis of the results of measurements of the temperature dependence of the integral intensity of X-ray scattering [36]; the data are averaged over several reflections. Note that the squared mean-square atomic displacements depend linearly on the temperature: for silicon to 1100 K and for germanium to 900 K.

The slopes of the lines characterize the forces of the bonds in the crystal lattice. From data obtained, in accordance with formula (3.3), the force constant α in the harmonic approximation can be calculated. The force constants of silicon and germanium are equal to 80.4 N/m and 56.4 N/m, respectively, and are essentially less than those of diamond. The reduced mean-square atomic displacement is in silicon 2.14×10^{-50} kg $\times$ m^2/K (in diamond, 0.39×10^{-50}). Figure 4.3 shows that the influence of increase of temperature on the crystal lattice of germanium is more marked than in silicon. Again, we see the role of the electron shells of atoms in the series of the elements (see Table 4.1): in germanium, the bond of the valency electrons with the nucleus is weaker than that in silicon.

The methods of X-ray and neutron diffraction analysis provide the possibilities of detailed and informative investigation of the nature of the covalent bond and of the problem of amplitudes of vibration of the valency electron shells providing this bond. The direct proof of the tetrahedral location of the covalent bonds in the crystal lattice

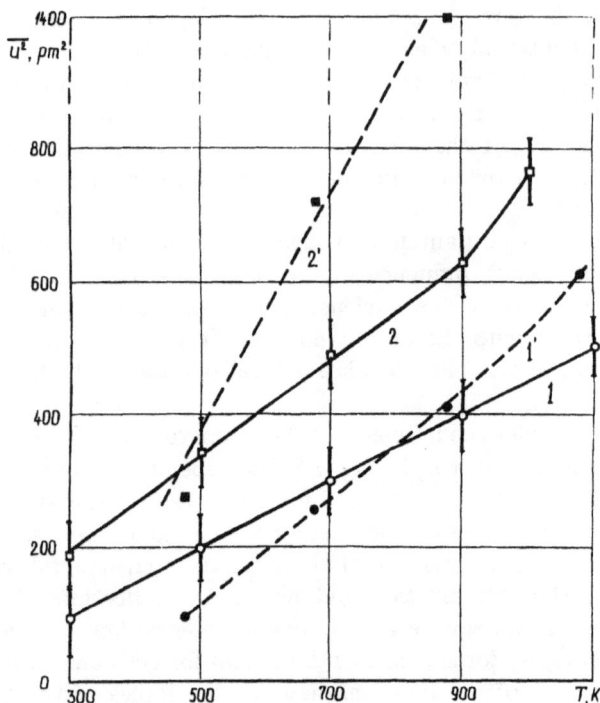

FIGURE 4.3. Influence of the temperature on the mean-square amplitude
of atomic vibrations in semiconductors:
1 and 1′ — silicon; 2 and 2′ — germanium; 1 and 2 are the average values
obtained from results of the measurements of "allowed" X-ray reflections
of the type (422), (333), (440), (531). 1′ and 2′ are the values obtained by
the data of intensity of the "forbidden" reflex (222).
The vertical intervals show the errors of the measurements.

of the diamond type is the presence of the so-called "forbidden" reflexes in the diffraction pattern. Among these are, for instance, the
X-ray reflection (222). As one of the factors in the expression for the
intensity of scattering (2.13), the structure factor is present. For the
X-ray reflection, it is equal to $F_{222} = 4(f_A - f_B)$, where f_A and f_B
are the factors of the atomic scattering of atoms which are in the A
and B positions. If the atomic shells had the spherical symmetry,
the difference of the atomic factors would be equal to zero and the
reflections of (222) should not be observed. In the general case, no

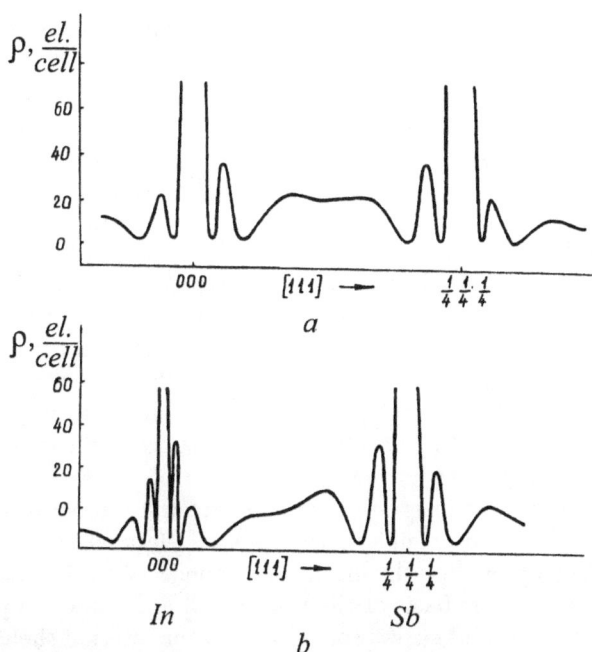

FIGURE 4.4. Distribution of the electron density along the direction [111] of the crystal lattice:
a — grey tin; b — the compound InSb [37].

reflections with the sum of indices $h + k + l = 4n + 2$, where $n = 0$, 1,..., are expected. However, in the experiments, the "forbidden" interferences, although weak but exceeding the limits of the possible error of the measurements, are observed. This can be explained uniquely. The covalent bond provides the tetrahedral redistribution of the valency electrons and they contribute to the intensity of the reflexes which should not be present. In other words, the presence of the weak reflections is a consequence of the distribution of the electron density along the interatomic bond (it is the electrons that scatter the X-ray radiation). This phenomenon is also detected when scattering the neutrons in silicon and germanium [38] and is studied in detail on the same materials in [39] as well as on grey tin [37].

In Fig. 4.4, the graph of distribution of the electron density along the direction <111> in a crystal of grey tin calculated in [37] is presented. In the central part of the graph, at an equal distance from two neighboring atoms, a noticeable growth of the electron density can be observed that is caused just by the cloud of electrons. This part of the graph naturally corresponds to the domain of addition of two overlapping wave functions of the type shown in Fig. 4.1.

With reference to the amplitude of the vibrations of covalent electron bonds, the opinions of the authors are different to a certain degree. Bilderback and Colella assume that, from their experimental data, the thermal displacements of the valency electrons in grey tin are the same as those of the shell electrons or, at least, are reduced by 20 %. The authors give the values at temperature 200 K and equal to 319 and 250 pm^2 for internal and covalent shells, respectively. Similar values are given in paper [40] for silicon. The authors studying Si and Ge [39] assume that the mean-square amplitudes of the vibrations are equal for covalent and internal electrons.

Therefore, it is interesting to compare the temperature dependence of the displacements of the internal electron shells of atoms and electrons providing the interatomic bonds. We calculated experimental data on the basis of the results [39] and these are presented in Fig. 4.3. The solid curves correspond to the internal shells and the dotted curves correspond to the covalent shells, and Fig. 4.3 shows that, for relatively low temperatures, the amplitudes for the first ones are greater than those for the second ones. However, the value $\overline{du^2}/dT$, which characterizes the variation of the amplitude with the temperature, conversely, is greater for the bonding electrons. Therefore, if the temperature exceeds 800 K for silicon and 500 K for germanium, then the amplitudes of the vibrations of electrons of valent bonds are essentially greater. For the temperature 1123 K (this point is absent in the graph), in germanium, their amplitudes makes 21 % of the interatomic distance.

Thus, we obtain the following result. It is known that the internal electron shells belonging in solids to the "quick" subsystem that vibrates with the same frequency and amplitude as the atomic nuclei belonging to the "slow" subsystem. The parameters of the vibrating motion of the paired covalent electrons are different and this difference depends on the temperature. For room temperatures, their amplitudes are somewhat less, but they essentially increase with growth of the temperature. One can say that, first of all, the heat motion loosens, shatters, shakes the interatomic bonds.

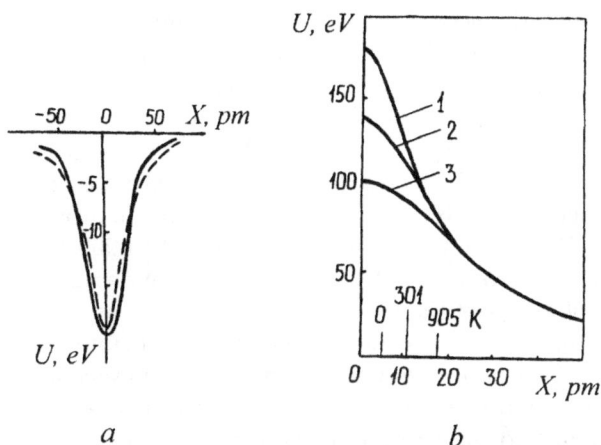

FIGURE 4.5. Dependence of the potential energy of interaction of the charged particles with the atomic channels on the distance [42, 43]: a — diamond; the interaction of electrons with the plane channel (110); the dotted line shows the heat vibrations of the lattice. b — the interaction of protons with the atomic row [110] in silicon. The temperatures: 1 — 0; 2 — 301; 3 — 905 K. The mean-square amplitudes of atomic vibrations in directions perpendicular to the atomic series are marked by vertical dashes.

Valuable information concerning the effect of the heat vibration on the interatomic interaction is given by the method of channeling of charged particles: electrons, positrons, and protons. The particles introduced in a monocrystal at a small angle to the atomic planes or rows exhibit an anomalous penetration and anomalously low losses of energy [41]. The motion of such particles is described by the laws of the classical physics, because the de Broglie wave-length is less than the interatomic distance by 3–4 orders. The method allows us to determine the potential energy of the interaction of the particle with the atoms of the plane or the rows in the material under investigation.

In Fig. 4.5a, the potential energy of the interaction of electrons with the plane channel (110) of a diamond crystal is shown [42]. The interplane distance is 253 pm; the figure shows that the efficient width of the channel is essentially smaller. Also, the heat motion weakens the interaction and washes out the channel increasing its efficient width.

Protons with energies in the range 3–11 MeV were used for the investigation of silicon and germanium monocrystals [43]. The fulfilled experiments and theoretical analysis made possible to establish the details of interaction of the particles with plane channels and atomic rows. This interaction is noticeably affected by the heat atomic vibrations. Fig. 4.5b illustrates the action of the elevated temperatures onto the energy of interaction of protons with the atomic row [110] of silicon. When the temperature rises from 0 to 905 K, the energy decreases from 175 to 100 eV; and the mean-square amplitude of the vibration of atoms of the row increases from 5 to 18 pm.

The atomic vibrations essentially affect the main — electrical — properties of classical widely used semiconductors such as silicon and germanium. In intrinsic pure semiconductors, the free charge carriers arise only due to the rupture of the covalent bonds. The heat displacements of the atoms result in gradual weakening of a part of the interatomic bonds, washing out the wave functions and generating free electrons, which are able to travel along the crystal lattice in the external applied electrical field. At the same time, the so-called holes are formed, i.e., unsaturated positively charged interatomic bonds which are also able to travel along the lattice. The process of heat liberation of the covalent electrons depends on the temperature and has a probabilistic character. Analyse this process using the Boltzmann distribution for the average number of ruptured bonds.

The number of ruptured interatomic bonds can be represented as

$$n = n_0 \exp\left(-\frac{U_0 - w}{kT}\right), \qquad (4.2)$$

where n_0 is the total number of covalent bonds per atom in a unit volume; U_0 is the energy of the process activation; w is the energy delivered by heat vibrations (both the energies are per atom); k is the Boltzmann constant; T is the temperature. Suppose in the first approximation that the energy of atomic vibrations can be expressed as the energy of harmonic oscillator: $w = m\overline{\omega^2}\dfrac{\overline{u^2}}{2}$. Substituting this expression into formula (4.2) and taking its logarithm, we obtain

$$\ln \frac{n}{n_0} = -\frac{U_0}{kT} + \frac{m\overline{\omega^2 u^2}}{2kT}. \qquad (4.3)$$

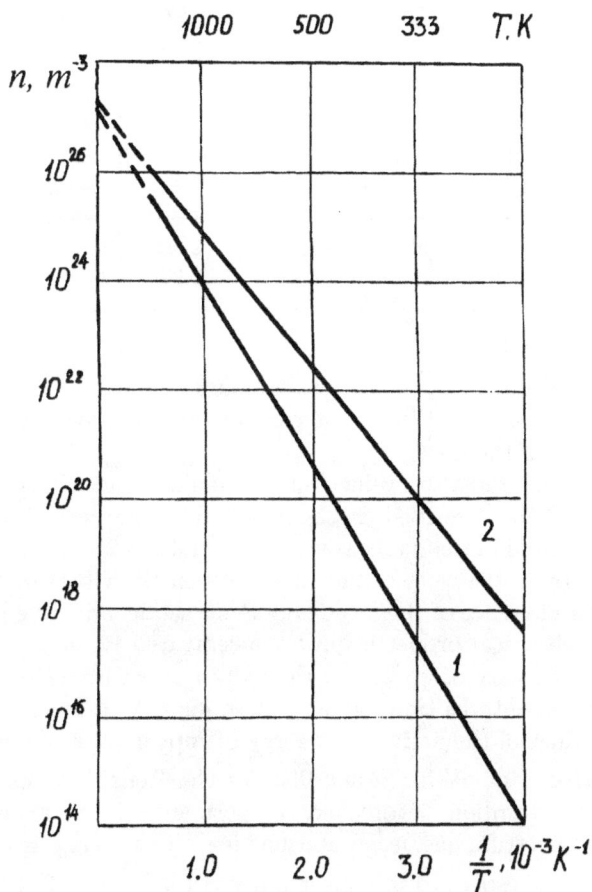

FIGURE 4.6. Dependence of the concentration of the charge carriers in semiconductors on the inverse temperature:
1 — silicon; 2 — germanium.

The experimental dependences of the concentration of the charge carriers — electrons — on the inverse temperature are, as should be expected, linear. They are given in Fig. 4.6.

The value $\overline{du^2}/dT$ is known from the experimental data, Fig. 4.3. One can see from formula (4.3) that, extrapolating the straight lines to $1/T \to 0$, we can determine the second term of the right-hand

TABLE 2. Activation energy of rupture of valency bonds and some parameters of the atomic vibrations in Si and Ge

Element	$\frac{\overline{du^2}}{dT}$, $\frac{pm^2}{K}$		$\frac{d\ln(n/n_0)}{d(1/T)}$, K	n_0, $10^{29}m^{-3}$	U_0, eV	$\sqrt{\overline{\omega^2}}$, 10^{13} rad/c
	internal shell	valency electrons				
Si	0.50	1.00	−7516	2.00	0.65	5.09
Ge	0.72	2.25	−5583	1.77	0.48	2.19

side and find the value of the mean-square angular frequency of the atomic vibrations $\overline{\omega^2}$. Finally, we can calculate the activation energy by the slope of the lines.

Table 2 contains the values for germanium and silicon obtained as a result of the described processing of the experimental data. The values presented in this table as well as formula (4.2) and the graph in Fig. 4.6 express the close connection between the vibration of atoms in the crystal lattice of the most important semiconductive materials and their electrical properties in a concentrated form. As has been noted, the velocity of growth of the values $\overline{u^2}$ with increase in temperature is two–three times greater than for the internal ones. Reasonable values of the activation energy of rupture of the valent bond are obtained: 0.65 eV for Si and 0.46 for Ge. Tenable values of $\sqrt{\overline{\omega^2}}$ attract the attention. Moreover, for germanium, the experimental frequency spectrum has been published [44]. The averaging of the frequencies with respect to this spectrum gives $\sqrt{\overline{\omega^2}} = 3.79 \cdot 10^{13}$ rad/s; the fit with the calculated value (Table 2) is satisfactory.

At this point, we finish the analysis of the process of the atomic vibrations in crystals of elements of the fourth group of the periodic table and the review of double compounds begins in the next subsection.

4.2. Semiconductive compounds of elements of the III and V groups

Double compounds of elements of the first subgroups of groups III and V of the periodic table with each other have a crystal lattice of the type of zinc blende. The nodes of type A are occupied by atoms of

TABLE 3. Some characteristics of initial elements and results of measurements of amplitudes of the atomic vibrations in compounds $A^{III}B^{V}$

	15 30.97 P $3s^2 3p^3$	33 74.92 As $4s^2 4p^3$	51 121.8 Sb $5s^2 5p^3$
31 69.72 Ga $4s^2 4p^1$	GaP 7.75 9.54	GaAs 9.75 9.22	GaSb 11.2 10.4
49 114.8 In $5s^2 5p^1$	InP 7.36	InAs 9.24	InSb 13.9 12.7 17.4 12.2

REMARK. The numbers in the cells of chemical elements denote: the number of the element; its atomic mass; the structure of the valency electron shell. The pairs of numbers in the cells of compounds $A^{III}B^{V}$ denote the values of the mean-square amplitudes in picometres for the atoms A and B, respectively. One number is the average value of the amplitude for the given compound.
According to data of [45, 47, 48].

an element of the III group and the B nodes, by atoms of an element of the V group. Typical representatives of this type of compound are GaP, GaAs, GaSb, InP, InAs, InSb.

Experimental determination of the amplitudes of atomic vibrations in semiconductive compounds of type $A^{III}B^{V}$ are described in several studies. Researchers working with monocrystals determined the values of the amplitudes separately for each of the two atomic sublattices. These data are most interesting. In other studies, the mean atomic displacements for a given material are measured.

In Table 3, the results of measurements for most wide-spread compounds of the considered type are compared. When they are formed, the valency bonds are provided by p-electrons; in this case, the element A supplies one electron and element B supplies three electrons. As the principal quantum number n of valency shells increases, the energy of bond of the electron with the nucleus decreases.

As should be expected, in this case, the value of $\sqrt{\overline{u^2}}$ increases appropriately. So, in the series of compounds GaP–GaAs–GaSb the displacement of the Ga-ion increases naturally from 7.75 to 11.20 pm. Comparing the amplitudes of the atomic vibrations in their sublattices, one can say that the more massive atoms displace by smaller distances. The value of the reduced amplitude is greater for atom A, except the compound GaSb. The absolute values of the amplitudes for GaAs and InAs are the same as for silicon; the data for GaSb and InSb are nearer to the results for germanium.

The data of intensity measurements of "forbidden" reflexes are the source of information concerning the nature of the covalent bond in semiconductors of the type $A^{III}B^{V}$. For instance, in the case of InSb, the reflection (600) exhibits an anomalous temperature dependence: its intensity grows as the temperature rises. The structural factor has the form

$$F^2_{(600)} = \left[f_{In} \exp(-M_{In}) - f_{Sb} \exp(-M_{Sb}) \right]^2. \qquad (4.4)$$

This experimental fact can be explained only by assuming that, as the temperature rises, the amplitude of the vibrations of the atom of indium grows essentially faster than the displacement amplitude of atoms of antimony. The mean-square amplitudes are present in exponents in expression (4.4); as the temperature rises, the first term of the sum decreases faster than the second one and the squared difference increases.

The difference between reflections (222) and $(\bar{2}\bar{2}\bar{2})$ indicates the nonsphericity of the valency electron shells. The results of measurements lead the authors of papers [45, 49] to the conclusion on a tetrahedral distortion of the density of the valency charge and on the transition of the charge from the atom of indium to the atom of antimony. Figure 4.4b shows that the cloud of valency electrons is displaced to the ion of antimony. This can be caused by the charge of the nucleus of antimony being greater than those of indium (51 versus 49). It seems that the displaced cloud of electrons enlarges the force constant for atoms of antimony and hinders their vibrations in a higher degree than for atoms of indium.

The authors of the cited papers gave two parameters of the model: one of them, α, determines the degree of the tetrahedral distortion of the bond (the parameter of "covalency"); the second one, γ, results from the transition of the electric charge from one atom to another one (the parameter of "ionicity"). Both α and γ are present in the equation for the distribution of the electron density. Both the parameters depend on the temperature: as it rises, the bond weakens. Similar data on the transition of the electric charge from Ga to As in compound GaAs are described in the literature [50].

Note, however, that there is an opposite point of view. The authors of the studies [51] believe that the transition of approximately 0.5 electrons per atom from B^V to A^{III} occurs. This conclusion is based on X-ray measurements of the atomic factors in the powders GaS, InAs, and GaSb, and on the comparison of their results with the data of the calculations on the base of the models. However, we think that the information obtained on the basis of direct measurements of the scattering of X-rays by valency electrons is more significant.

One of the most important characteristics of a semiconductive material, namely, the width of the forbidden band E_g, is the value of the energy gap between the ceiling of the valent band and the bottom of the conduction band. In particular, the value of E_g should be provided by the energy of the interaction of neighboring atoms. The energy of the interaction also affects the amplitude of the heat displacements, and this leads us to conclude that the two quantities correlate.

On the graph in Fig. 4.7, the values of the width of the forbidden gap for a series of semiconductors (determined by the position of the edge of the fundamental absorption of the light quanta) and the corresponding values of the reduced amplitude of the atomic vibrations are shown. The inverse correlation between these two characteristics of the semiconductors is clearly verified. The growth of the reduced amplitude corresponds to the decrease of the width of the forbidden gap.

It is interesting that the semiconductive materials under investigation can be naturally divided into four groups, taking again the structure of the electron shells and the appropriate intervals of the variation of the parameters as the criteria. Note the data in Table 4, which shows this division and the sum of the principal quantum numbers of the valency electron shells increases from the first group to the fourth group. Accordingly, the interatomic bonds weaken, the quantities η grow and the values of E_g decrease. Of course, the sum $n_A + n_B$ is a conditional and qualitative measure of the strength of the covalent bond. However, the emphasized simple correlation is of

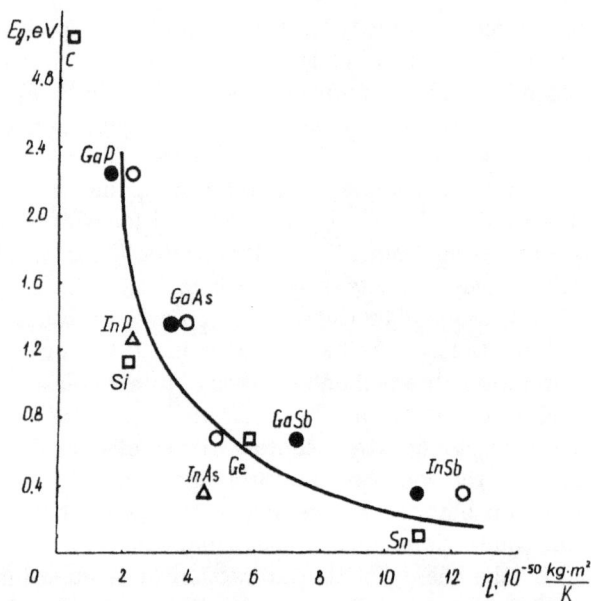

FIGURE 4.7. Correlation between the width of forbidden band in semiconductors and the values of the reduced amplitude of atomic vibrations.

TABLE 4. Valency electron shells and some characteristics of semiconductors

Group	Forming of electron shells	$n_A + n_B$	$\eta,$ 10^{-50} kg·m^2/K	$E_g,$ eV
C	$2s^1 2p^3$	4	< 2.0	> 2.2
GaP	$4p^1 + 3p^3$	7		
Si	$3s^1 3p^3$	6	2.0–4.0	1.1–1.4
InP	$5p^1 + 3p^3$	8		
GaAs	$4p^1 + 4p^3$	8		
Ge	$4s^1 4p^3$	8	4.5–7.5	0.4–0.7
InAs	$5p^1 + 4p^3$	9		
GaSb	$4p^1 + 5p^3$	9		
Sn	$5s^1 5p^3$	10	> 10.0	< 0.4
InSb	$5p^1 + 5p^3$	10		

interest. It is the nature of the interatomic bond that determines the main properties of the semiconductive compounds widely used in practice.

4.3. Semiconductive compounds of elements of groups II and VI

The compounds of the type $A^{II}B^{VI}$ attract the attention of researchers because of their connection with practical applications as well as the ambition to the study of the fundamental laws of interatomic bonds. These chemical compounds have semiconductive properties, in many ways similar to $A^{III}B^{V}$; however, they have peculiarities and distinctive features. The materials based on them are applied in developing areas of microelectronics, for example, in the manufacture of multilayer structures (superlattices).

The interatomic bond in compounds of this class is formed by valency electron shells n_1s^2 (atom A) and $n_1s^2n_2p^4$ (atom B). (Here, n_1 and n_2 are the quantum numbers, $n_1 < n_2$.) This structure of the initial electron shells of the atoms of elements forming the bond results in the shift of the negative electric charge from the ion A to the ion B. As a result, an essential ionic component of the interatomic bond appears. Thus, in the compounds under consideration the interatomic bond is by its nature partially covalent and partially ionic.

Typical representatives of the chemical compounds of this class are ZnO, ZnS, ZnSe, ZnTe, CdO, CdS, CdTe.

The compound CdTe has a crystal lattice of the type of zinc blende. In Fig. 4.8, the dependences of the mean-square amplitudes of the atomic vibrations in cadmium telluride on the temperature measured in the direction [100] are presented. The graphs are constructed on the basis of the data of experimental investigation of a monocrystal using X-ray scattering [52]. Comparing CdTe with the similar compound InSb (the differences of the atomic masses are quite small), note that the mean-square amplitudes for the first compound are slightly greater. This demonstrates that the interatomic bond is weakened. The dependences $\overline{u^2}(T)$ are linear in the studied temperature interval and, as the temperature decreases, they extrapolate to zero. The more massive tellurium atom attracts a part of the valency atomic charge and vibrates with a lower amplitude (curve 2 in Fig. 4.8). The values of $d\overline{u^2}/dT$ are greater for the cadmium atom.

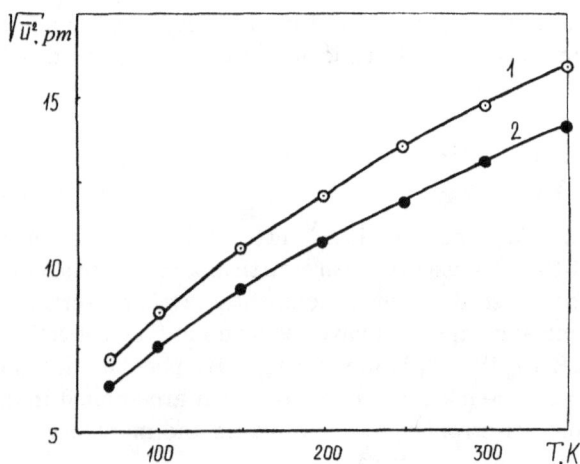

FIGURE 4.8. The effect of temperature on the mean-square amplitudes of atoms vibrations in the compound CdTe [52]:
1 — atom of Cd; 2 — atom of Te.

Just as in the case of InSb, in a number of ionic compounds, the reflexes of types (200), (600), and (10 00) reveal an anomalous temperature dependence: their integral intensity increases as the temperature rises. We have considered the nature of this phenomenon in the previous subsection; in this case, they are caused by the faster growth of the vibration amplitude of the atom Cd compared with those of the atom Te.

In the compound CdTe, the effects of anharmonicity are essential; the potential "well", in the limits of which the vibrating cadmium atom deviates, has less sloping walls than the tellurium atom. We shall return to this problem in the last section.

According to the published data of the neutron-structural analysis [53], the values of the vibration amplitudes of the atoms in ZnTe for room temperature are 15.7 and 12.4 pm, respectively. We see that they differ from the parameters of CdTe within the measurement errors. Again, the amplitude of the atom A is greater. The average atomic displacements in the compounds ZnO, ZnS, and ZnSe are within the limits 22.0 – 22.2 pm [54]. The oxide ZnO, contrary to the two other mentioned compounds, has a hexagonal crystal lattice; and a noticeable anisotropy of heat vibrations is typical. The anisotropy of the heat displacements is also observed in the compounds CdS and CdSe, which have the wurtzite structure.

4.4. Ionic compounds $A^I B^{VII}$

It is known that the nature of the ionic interatomic bond is electrostatic: the electron passes from one atom to another and the Coulomb interaction arises between the ions.

While considering the results of the study of the ionic compounds of the first group metals of the periodic table with halogens, we note the relatively large values of the vibration amplitudes of the metallic atom. For instance, in the compound CuJ, the amplitude is equal to 90 pm at temperature 673 K [55]. This exceeds by several times the values, which we have met in the previous subsection. In experiments, as the temperature rises, the known for us growth of the intensity of reflexes, for which the structural factor is defined by the difference of the factors of atomic scattering of the elements, is observed. For ionic compounds, as has been established, the essential anharmonicity is typical. This is the reason, why, when expanding the potential energy of the interatomic interaction into a series, it is necessary to take into account the terms of higher order than quadratic one. A number of studies are devoted to the determination of the force constant β (see Section 6). In a study devoted to the neutron-structural investigation of copper chloride, it has been found that the copper atom performs asymmetric anharmonic vibrations in the direction of the interatomic bonds [56]. A similar phenomenon is also typical for other ionic compounds: CuBr, CuJ, and CaF_2.

Table 5 is devoted to the parameters of the atomic vibrations in ionic compounds. Unfortunately, in most papers, for these compounds, only the measurements of the average values of the mean-square amplitudes are performed. However, in the papers, where the separate determination of these values for the cation and anion is performed, the greater amplitude of displacements of the metal atom over those of the halogen atom is emphasized.

By and large the experimental material illustrates sufficiently clearly the growth of the amplitudes in the series $A^{IV} \rightarrow A^{III} B^V \rightarrow A^{II} B^{VI} \rightarrow A^I B^{VII}$.

TABLE 5. Amplitudes of atomic vibrations in ionic compounds of the type
$A^I B^{VII}$

Compound	$\sqrt{u_A^2}, \quad \sqrt{u_B^2},$ or $\sqrt{u_{av}^2},$ pm	Source
CuCl	40.9 30.2 42.0 30.0	[56] [57]
CsBr CsJ TlCl TlBr	29.7 31.4 32.5 32.1	[58]
$TlCl_{0.7}Br_{0.3}$ AgCl	33.3 28.5	[59] [60]

4.5. Carbides

The compounds of metals with carbon represent the interstitial phases.
An interstitial phase has a crystal lattice of metal, whose internodes
are occupied by atoms of carbon which are small in size; in this case,
the placement of the injected atoms in the metallic lattice does not
cause the change of its symmetry or results in small distortions of the
lattice.

The interstitial phases play an important role in engineering.
Many of them are the structural components of commercial steels
and alloys, are widely used in powder metallurgy. They provide the
valuable, unique service characteristics of materials such as the hardness, wear resistance, heat-resistance. It is natural that the choice
of objects for investigation of the lattice dynamics among this large
class of materials is determined by their importance for engineering
applications.

We begin the consideration of heat atomic vibrations with vanadium carbide VC, which has a cubic crystal lattice of the NaCl type.
By its chemical composition, it can often deviate from the stoichiometric content of elements and crystallize with deficiency in carbon,
forming a continuous series of solid solutions in the interval VC –
V_4C_3. It is also essential that, in complex multicomponent systems,
carbide can dissolve atoms of different metallic elements: Cr, W, Mo

TABLE 6. Composition of steels (the remainder is iron), formulae of carbides, amplitudes of atomic vibrations at 873 K and temperatures of loss of strength of steels to a hardness of 40 HRC.

	Composition	Carbide formula	$\sqrt{u_M^2}$, pm	T, K
1	0.7C–4V–2Cr	$(V_{0.93}Cr_{0.07})C_{0.80}$	18.5	915
2	0.7C–4V	$V_{1.00}C_{0.82}$	17.1	903
3	0.7C–2V–1W	$(V_{0.92}W_{0.08})C_{0.96}$	15.8	921
4	0.4C–2V–3W	$(V_{0.89}Fe_{0.02}W_{0.09})C_{0.88}$	15.2	938
5	0.4C–2V–2Cr	$(V_{0.81}Fe_{0.08}Cr_{0.11})C_{0.77}$	18.7	907
6	0.4C–2V–1Mo	$(V_{0.87}Fe_{0.04}Mo_{0.09})C_{0.82}$	16.0	930
7	0.4C–2V–2Mo	$(V_{0.80}Fe_{0.02}Mo_{0.18})C_{0.80}$	14.8	935
8	1.0C–2V–4Cr –2Si–1Mo–2W	$(V_{0.64}Fe_{0.06}Cr_{0.08}W_{0.06}Mo_{0.10})C_{1.00}$	13.8	943
9	0.4C–2V–2Cr –1Si–2W–1Mo	$(V_{0.62}Fe_{0.06}Cr_{0.04}W_{0.08}Mo_{0.10})C_{1.00}$	14.5	943

and so on, forming the phases of variable composition with a wide domain of homogeneity. A carbide on the base of VC plays the role of the main strengthening phase in many steels which are employed for production of cutting and pressing tools.

The author together with his co-workers investigated the influence of temperature and the chemical composition of the carbide on the base of VC on the value of the mean-square amplitudes of atomic vibrations in both sublattices. The preparation of the material for the investigation was intended to obtain the carbide phases alloyed so that the influence of different elements on the heat vibration can be traced. Steels of different marks were melted out and thier compositions were such that, mainly, the carbide was formed with the crystal lattice on the base MC – M_4C_3 (M is a conditional notation of the metal atom). Among others, industrial steels of two different compositions were taken intended for tools of cold and hot deformation, respectively; their compositions are shown in last two lines of Table 6. The extraction of carbide phases was performed by the electrolytic method from specimens, which were cut from forged rods. The obtained precipitate was used for X-ray-diffractometer measurements of the amplitude of atomic displacements at high temperatures; afterwards, this precipitate was subjected to chemical analysis. The details of the procedure are described in [13]. The method of separate determination of the value of atomic displacements in metal and carbide sublattices is described in the second section of this review.

The results of determination of the chemical formulae of carbide and the mean-square amplitudes are presented in Table 6. The stoichiometric composition of carbide is not observed in all studied specimens; its chemical formula is close to M_4C_3; the exception is the carbide of industrial steels. In the crystal lattice of the carbide, the octahedral internodes, which are not occupied by carbon atoms, are vacant. Thanks to the obligate variation of the degree of alloying of the steels, we managed to obtain carbide phases with different content of chromium or manganese, molybdenum, tungsten in the solid solution. As a rule, a small part of atoms of the main element of steel, iron, is also present in the composition of carbide.

The carbide $VC_{0.82}$ extracted from the steel 0.7C–4V is basic, and the mean-square amplitude of vibration of the metal atoms is equal to 17.1 pm, and the force constant is 75.0 N/m. Introduction of chromium or manganese atoms to the composition of carbide increases the amplitudes of the vibrations of the metal atoms and at the same time decreases the value of the force constant.

As a result of alloying the carbide by molybdenum and tungsten, the interatomic interaction grows. Atoms of these elements replace vanadium atoms and promote the decrease of the mean-square amplitudes. The smallest values of the amplitudes are obtained for carbides of complex composition extracted from the steels of industrial qualities. Besides the chromium and iron atoms, carbides also contain tungsten and molybdenum atoms in the metal sublattice. For instance, the value of $\sqrt{u_M^2}$ decreases to 13.8 pm, and the value of α grows to 91.8 N/m. The analysis of experimental data shows that the variation of the amplitude of atomic vibrations is not proportional to the growth of the average atomic mass. Figs. 4.9 and 4.10 show the effect of the temperature on the amplitudes of atoms vibrations in the metal and carbon sublattices of carbide. It is interesting that the carbon atoms vibrate with amplitudes in 2–3 times greater than those of metal atoms. At the same time, the average atom masses differ more essentially, by four–six times. Obviously, it is caused not by

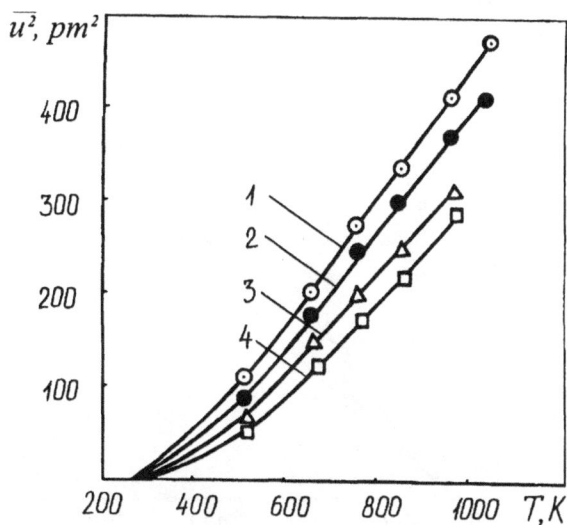

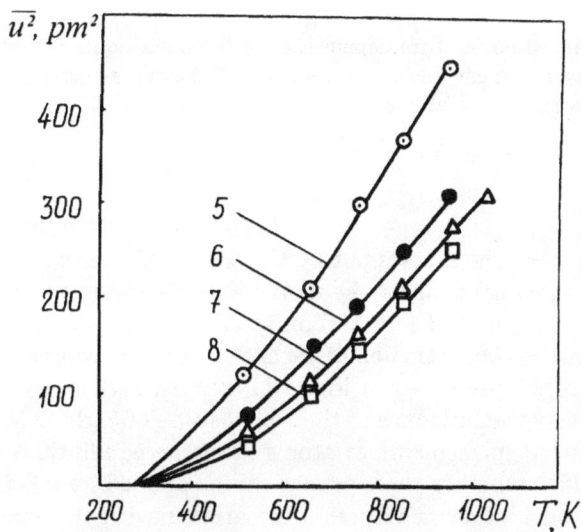

FIGURE 4.9. Temperature dependence of the mean-square displacements of metal atoms in carbides extracted from the studied steels. The numbers of the curves correspond to the steel compositions given in Table 6.

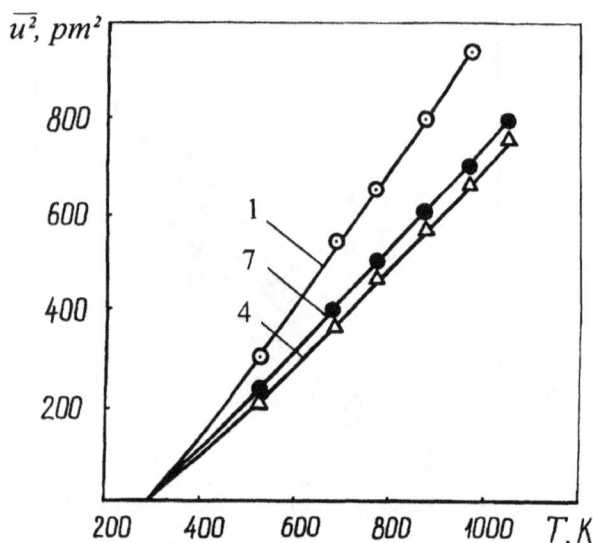

FIGURE 4.10. Temperature dependence of the mean-square displacements of carbon atoms in carbides. The numbers of the curves correspond to the steel compositions in Table 6.

the difference of the masses but by the energy of interatomic interaction. The alloying of carbide by molybdenum and tungsten decreases the vibration amplitudes of carbon atoms; the second reason is the decrease of the concentration of vacancies in the carbon sublattice.

The graphs presented in Fig. 4.11 were constructed to estimate quantitatively the influence of the composition of carbide MC on the mean-square displacements of atoms from the equilibrium position. It was verified whether these graphs can be applied for calculation of the amplitudes of atom vibration in carbides with the known composition assuming that every element affects independently, i.e., the contribution into the displacement is additive (for rather small concentrations of alloying elements, this assumption is reasonable).

For example, a satisfactory agreement of calculated values of $\overline{u^2}$ with the experimental ones is obtained for the carbides in industrial steels.

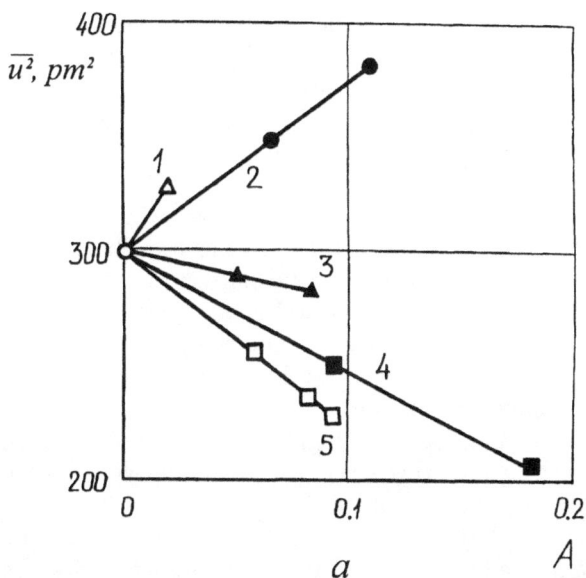

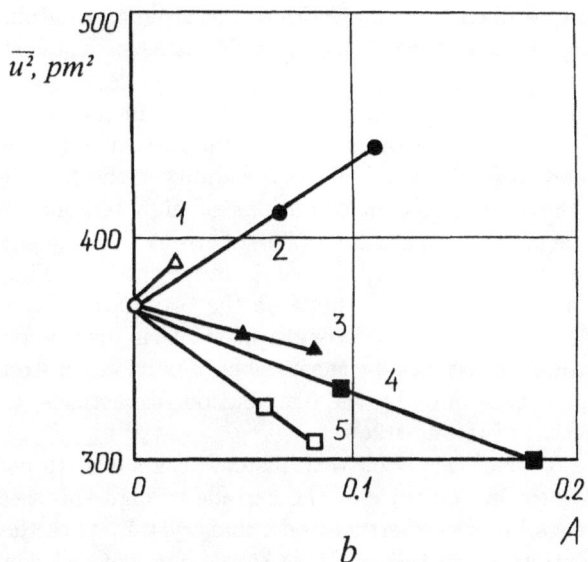

FIGURE 4.11. Affect of the composition of carbide MC–M$_4$C$_3$ on the mean-square vibration amplitudes of metal atoms:
A — atomic parts of admixtures: 1 — Mn; 2 — Cr; 3 — Fe; 4 — Mo; 5 —W. Temperature: a — 873; b — 973 K.

Figure 4.11 demonstrates the effect of any investigated element on the vibration amplitudes and, accordingly, on the forces of bonds between atoms M–M. The presence of chromium and manganese atoms in carbide diminishes these forces. The dissolution of the molybdenum and, especially, tungsten atoms in carbide has the opposite effect.

Note that the obtained data can be used while constructing new steels of high-performance qualities. In this case, both the properties are important: so, to dissolve the carbides in austenite while heating for hardening, it is sometimes necessary to weaken the forces of bonds between the atoms; conversely, during the period of steel operation it is important to strengthen the interatomic interaction and, thus, to slow the diffusion processes.

The last column of Table 6 gives the values of the temperatures of loss of strength of this steel, i.e., the temperature to which the steel should be heated in order that its Rockwell hardness decreases to 40 units. The higher is this temperature, the more thermostable is the steel and the better are its operation characteristics. The inverse correlation between the amplitudes of the atomic vibrations and the temperature of loss of strength can be clearly seen. The lowest values of the amplitudes and, accordingly, the highest temperatures of loss of strength are typical for industrial steels, in which carbide is alloyed complexly. The increase of the forces of interatomic bonds diminishes the diffusion coefficients and affects the ability of the carbide particles to resist the coagulation in conditions of high temperatures. The dependence of the temperature of loss of strength of the steels on the amplitude of the atomic vibrations is represented in Fig. 4.12 and demonstrates the correlation between the two quantities. Of course, the temperature of loss of strength of steels is also determined by other factors, for instance, by the number of particles of strengthening phase per volume unit, by the distribution of particles, and by the thin structure of the matrix.

The author of this book with his coworkers has studied another carbide known in literature as the carbide of high-speed steel. This carbide has a intricately-structured cubic crystal lattice that belongs to the spatial group $Fd3m$ [61]. An elementary cell contains 112 atoms, among them 96 metal ones and 16 carbon ones. The cell can be represented as consisting of octahedra and tetrahedra; at the vertices of octahedra, the atoms W or Mo are located, and at the vertices

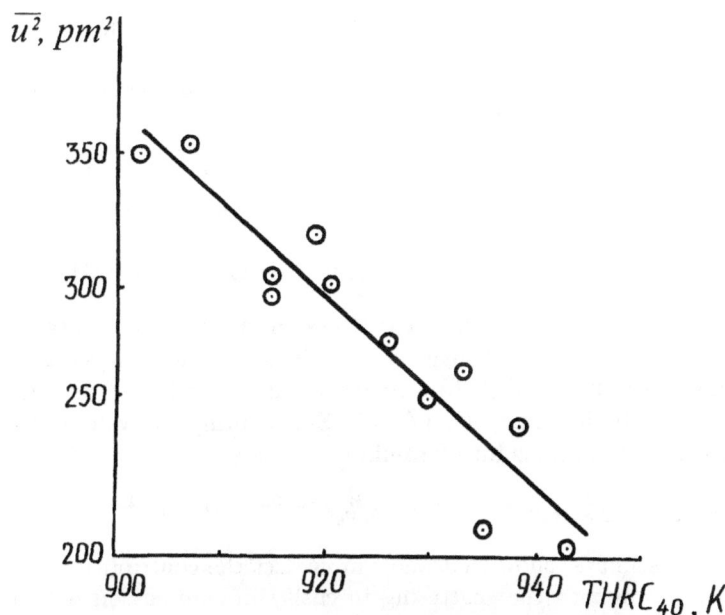

FIGURE 4.12. Correlation between the vibration amplitudes of the metal atoms in carbides at 973 K and the temperature of loss of strength of the tool steels.

of tetrahedra, the atoms Fe. The carbon atoms are located halfway between two octahedra and each is circled by six metal atoms [62]. In other words, 48 tungsten or molybdenum atoms occupy the positions (f); 48 iron atoms, positions (d); 16 carbon atoms, positions (c). Thus, the complete formula of the carbide must be $Fe_{48}(W,Mo)_{48}C_{16}$. In the real metallic systems, the considered complex carbide represents a phase of variable composition existing in a wide domain of homogeneity. Carbide is an important structural component of high-speed steels and essentially affects their properties. The steels applied for production of cutting tools are very widely used in industry, and the optimization of the content of costly and deficit alloying elements in these steels is an urgent problem.

The complicated structure of the carbide requires the analysis of the value of the structural factor and the choice of the interferences for investigation. Moreover, some simplifying assumptions should be inevitably taken.

The square of the structural factor of intensity of X-ray scattering for this carbide should be written:

$$F_{hkl}^2 = \sum_{r=1}^{112} \left(f_r e^{-M_r}\right)^2 \left[\cos^2 2\pi(hx_r + ky_r + lz_r)\right.$$
$$\left. + \sin^2 2\pi(hx_r + ky_r + lz_r)\right]. \quad (4.5)$$

All the notations in this formula are the same as in the second section. It turns out that, for the investigation, it is expedient to choose the interferences (400) and (800) that represent two orders of reflections from one and the same plane (100). Substituting the values of the basis for all 112 atoms into formula (4.5), we obtain

$$F_{hkl}^2 = 48 f_{W,Mo}^2 e^{-2M_{W,Mo}} + 48 f_{Fe}^2 e^{-2M_{Fe}} + 16 f_C^2 e^{-2M_C}. \quad (4.6)$$

The estimates show that one can neglect the contribution of the carbon atoms into the scattering intensity in comparison with the contribution of the metal atoms. Actually, the ratio $\dfrac{f_C^2}{4f_{Fe}^2} = 0.011$. On the right-hand side of equality (4.6), two unknown quantities $\overline{u_{W,Mo}^2}$ and $\overline{u_{Fe}^2}$ are present; strictly speaking, they are not equal to each other. Obviously, it is impossible to determine the amplitudes of displacements of every metal atom separately in so complicated structure. Therefore, we assume that $\overline{u_{W,Mo}^2} \cong \overline{u_{Fe}^2}$. Then we have

$$F_{hkl}^2 = 48\left(f_{W,Mo}^2 + f_{Fe}^2\right) \exp\left(-\frac{16\pi^2 \overline{u^2} \sin^2 \theta}{3\lambda^2}\right), \quad (4.7)$$

where $\overline{u^2}$ takes the meaning of the average value of the squared mean-square amplitude of the metal atoms vibrations in the crystal lattice of the type $M_{96}C_{16}$.

The intensity of the studied interferences can be written as

$$I = I_0 C A L(\theta) F_{hkl}^2, \quad (4.8)$$

where I_0 is the intensity of the primary ray; C is the product of the electron factors of intensity; A is the absorption factor; $L(\theta)$ is the product of the polarization factor and the Lorentz factor.

TABLE 7. Amplitudes of vibrations of metal atoms in crystal lattice of carbides of high-speed steels, pm

Composition of steel	293	673	773	873	973 K
1 0.8C–4Cr–1V–18W–1Mo	35.0	33.6	39.9	41.5	43.3
2 1.0C–4Cr–2V–12W–3Mo–5Co	33.1	38.2	39.5	40.2	41.1
3 1.0C–4Cr–2V–6W–5Mo–5Co	25.0	31.1	33.4	35.5	40.4
4 1.0C–4Cr–2V–2W–8Mo–5Co	17.2	25.5	28.8	31.0	31.0

Dividing equation (4.8) by $L(\theta)$ and taking its logarithm, with due account of (4.7), we obtain

$$\ln \frac{I}{L(\theta)} = \ln \left[I_0 \cdot C \cdot A \cdot 48 \left(f_{W,Mo}^2 + f_{Fe}^2 \right) \right] - \frac{16}{3} \pi^2 \overline{u^2} \frac{\sin^2 \theta}{\lambda^2}. \quad (4.9)$$

Thus, constructing the dependence of $\ln[I/L(\theta)]$ on $\sin^2 \theta / \lambda^2$ by the experimental data, $\overline{u^2}$ can be determined by the slope of the straight line; the interval cut on the y-axis should contain some information concerning the studied carbide, since its length depends on the factors of atomic scattering and absorption. The carbide composition affects both these factors.

The object of the study is the four carbide phases obtained by the method of electrolytic extraction from four steels. The composition of the studied high-speed steels is given in the first column of Table 7. The first of the listed steels is the classical steel that was before widely used for fabrication of the cutting tools; in the modern classification, it belongs to the steels of normal productive capacity. The three last steels additionally alloyed by cobalt belong to the steels of enhanced capacities. From the quality composition, one can see the different content of the currently deficit tungsten in the steels.

In Fig. 4.13, the experimentally obtained dependences are presented for carbides studied at the temperatures 673 and 873 K. For different carbide phases, different slopes of the curves and their different positions are typical. This is caused by the variation of the mean-square displacements of atoms and the carbide composition.

The results of processing the experimental data can be seen in Table 7. Table 8 contains the data obtained as a result of chemical analysis and X-ray diffraction study. As the concentration of tungsten

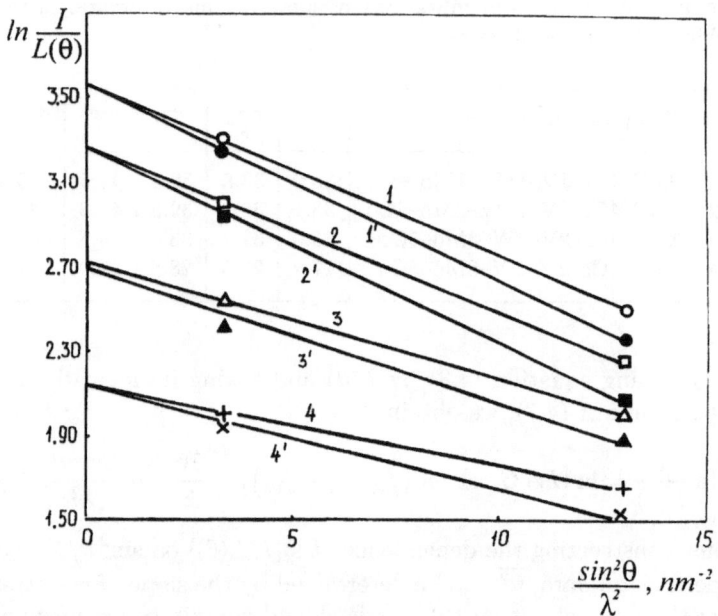

FIGURE 4.13. Dependence of $\ln[I/L(\theta)]$ on $\sin^2\theta/\lambda^2$ for carbide phases of high-speed steels. The number of curves correspond to data of Table 7.
1, 2, 3, 4 — $T = 673$ K;
1′, 2′, 3′, 4′ — $T = 873$ K.

TABLE 8. The composition of carbide phases of high-speed steels and some their characteristics

Steel	Formula of carbide	$f^2_{\text{W,Mo}}+$ $+f^2_{\text{Fe}}$	$\bar{\mu},$ cm^2/g	k
1	$[(\text{Fe}_{42}\text{Cr}_2)(\text{W}_{47}\text{Mo}_5)]\text{C}_{16-x}$	5701	2836	3.55
2	$[(\text{Fe}_{47}\text{Co}_2)(\text{W}_{32}\text{Mo}_{10}\text{V}_2\text{Cr}_3)]\text{C}_{16-x}$	4501	2618	3.26
3	$[(\text{Fe}_{40}\text{Co}_5\text{Cr}_2)(\text{W}_{24}\text{Mo}_{23}\text{V}_1)]\text{C}_{16-x}$	3995	2522	2.70
4	$[(\text{Fe}_{44}\text{Co}_3\text{Cr}_1)(\text{W}_5\text{Mo}_{35}\text{V}_1\text{Fe}_8)]\text{C}_{16-x}$	2909	2268	2.15

μ is the mass coefficient of X-ray absorption.

in steel decreases, a part of the tungsten atoms in carbide is replaced by the molybdenum atoms. Since the factor of atomic scattering of molybdenum is smaller than that of tungsten, the value of k is equal to the first term on the right-hand side of formula (4.9) decreases and the curves in Fig. 4.13 go down to the x-axis. It is interesting that, as the atomic part of molybdenum in carbide increases, the value of the vibration amplitude of the metal atoms in the complex lattice of high-speed steels noticeably decreases. The carbide particles become more stable against the coagulation in operating conditions of tool steels. Actually, the practice of employment of the studied steels confirms this conclusion.

Note that the mean-square amplitudes of metal atoms vibrations in carbide of the type M_6C exceed the values of amplitudes, typical for carbide MC, by almost twofold.

5. FREQUENCY OSCILLATORY SPECTRUM OF SOLIDS AND INTERATOMIC INTERACTION

5.1. Experimental dispersion curves and atomic force constants of metal elements

Methods of neutron spectroscopy based on the phenomenon of scattering of heat neutrons by crystals enable us to construct the dispersion curves of real solids and allow us to study the problems of physical nature of the interatomic interaction.

In Fig. 5.1, the experimental points of dependences of angular frequency of vibrations in aluminium on the wave vector are presented [63]. A particular presentation of experimental data is used: the abscissa is not the value of the wave vector itself, but the dimensionless quantity $\zeta = q/q_m$, which is equal to the ratio of the wave vector to its maximal value in the given crystallographic direction. The measurements of the frequency of atomic vibrations on monocrystals are first of all made in the directions of high symmetry of the type [100], [110], [111], as well as in some others. In the general form, the directions of vibrations are usually denoted $[\zeta 00]$, $[\zeta \zeta 0]$, $[\zeta \zeta \zeta]$ etc.

With reference to the typical data in Fig. 5.1a, the frequencies of the longitudinal L and transverse T oscillatory modes were measured on aluminium monocrystals in the direction of the side of the cubic elementary cell [100]. The experimental dependences resemble the dispersion curves for a linear atomic chain obtained by simple reasoning and calculations (Subsection 1.2 of Section 1). The curves begin as straight lines and their slope at the point 0, the origin, numerically corresponds to the velocities of longitudinal and transverse sound waves in aluminium computed on the basis of elastic constants

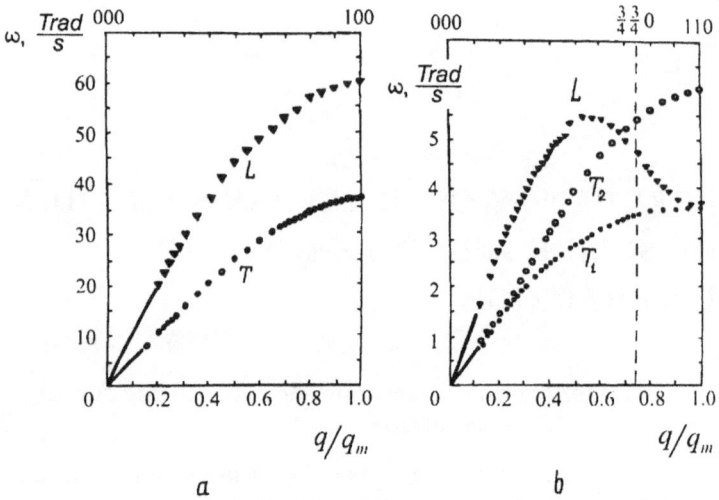

FIGURE 5.1. Experimental dispersion curves of atomic vibrations in the crystal lattice of aluminium. The temperature is 300 K. The x-axis corresponds to the dimensionless ratio of the value of the wave vector to its maximal value; the y-axis corresponds to the vibrations frequencies: a — the direction [100]; b — [110]; L is the longitudinal mode of vibrations; T, T_1, and T_2 are the transverse modes.

of aluminium. For the same value of the wave vector, all cyclic frequencies ω for the longitudinal mode of vibrations $L[100]$ are greater than for the transverse; the wave velocity is also greater for the longitudinal waves.

The graphs imply that, as soon as the wave length of the vibrations diminishes so that it becomes comparable with the interatomic distance, the curves deviate from straight lines. Accordingly, the group velocity of the elastic waves $d\omega/dq$ drops and at $q = q_m$ vanishes, at this point the tangent to every curve is parallel to the abscissa. The least interatomic distance in the cubic face-centred lattice of aluminium is equal to $a\sqrt{2}/2$ (a is the lattice period) and corresponds to the direction [110]. In this direction, three branches of the dispersion curve are obtained: one longitudinal L and two transverse: T_1 and T_2 (Fig. 5.1b). It is interesting to observe the longitudinal wave: as the ratio q/q_m grows, the wave velocity decreases and at

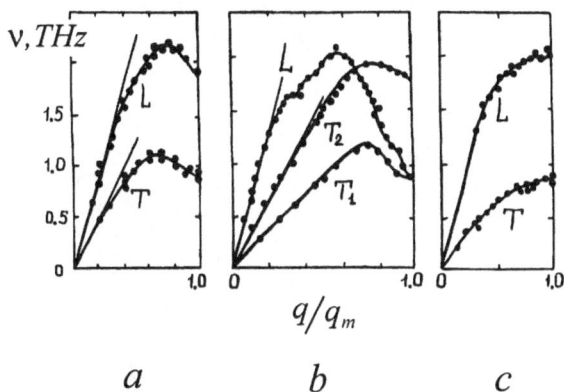

FIGURE 5.2. Dispersion curves for three directions of the propagation of the atomic vibrations in lead at the temperature 100 K [14]:
a — the direction $[\zeta 00]$; b — $[\zeta\zeta 0]$; c — $[\zeta\zeta\zeta]$.

$q = 0.5 q_m$ vanishes. The atom at the centre of the face is located just on the distance $a\sqrt{2}/2$ from the initial atom. Hence, for these values of the wave vector, the atoms at the vertices of the elementary cell and the atoms at the centres of the faces move in antiphase toward each other. Below, we consider this situation in more detail.

When the wave length continues to decrease, the velocity changes sign (the tangent to the curve makes an obtuse angle with the abscissa); the wave moves in the opposite direction. The wave of the T_2 mode is polarized in direction [001]; and of the T_1 mode, in direction [1$\bar{1}$0].

The high degree of symmetry of the cubic crystal lattice results in the fact that the vibrations of different types coincide at certain points. For instance, at the node [[110]] of the elementary cell, the oscillatory mode $L[110] = T_1[1\bar{1}0]$. In other words, the displacement of the atom along the diagonal of the face of the cube for longitudinal vibrations are similar to those occurring when the transverse wave propagates along the same diagonal. This phenomenon is called the degeneration of the oscillatory modes.

Dispersion curves obtained for lead monocrystals [14] are presented in Fig. 5.2. Qualitatively, the curves are similar to the dependence observed for aluminium, however, quantitative differences are obvious. For instance, the curves $L[\zeta 00]$ and $T[\zeta 00]$ pass through the

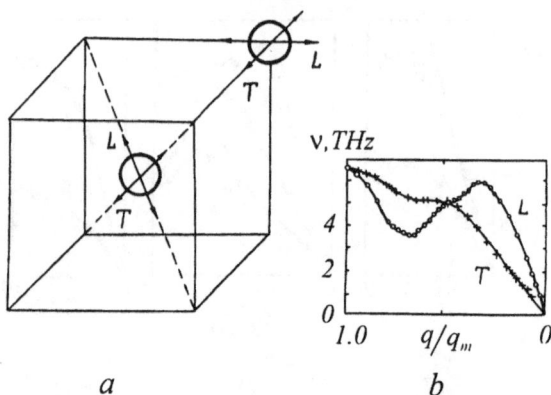

FIGURE 5.3. Some directions of vibrations in the cubic body-centred lattice
of niobium:

a — the scheme of longitudinal and transverse vibrations in the direction
[111]; *b* — the dispersion curves for the direction [111]; *L* and *T* are the
longitudinal and transverse modes, respectively.

maximum in the case of lead, but this is not valid for aluminium. Figure 5.2 shows that the abscissa of the maximum is $\zeta = 0,67$; hence,
the corresponding wave length is $\lambda = 3b$ (b is the least interatomic
distance in direction [100]).

There are inflection points on the curves $\nu(q/q_m)$ (see, for instance, the branch $L[\zeta\zeta0]$). These inflections are called the Cohn
anomalies; they appear as a result of the interaction of vibration
quanta, phonons, with electrons.

The degeneration of transverse and longitudinal oscillatory modes
at the centre of the cubic body-centred elementary cell of niobium is
illustrated by Fig. 5.3a. Actually, by virtue of the lattice symmetry
in the direction of the side of the cube, the frequencies of vibrations
$L[100]$ and $T[001]$ must coincide. The same figure also shows that, in
the direction of the spatial diagonal, the same vibration frequency of
the atom at the centre of the cell should be typical for the longitudinal
waves, as well as for the transverse ones. Accordingly, in Fig. 5.3b,
the curves L and T cross each other at the point corresponding to
$q/q_m = 0.5$ and to the frequency 5.1 THz (data of [64] for niobium).

Table 1 presents, as an example, the information concerning the
number of atoms in the cubic crystal lattice, which surround the atom

TABLE 1. Neighbouring atoms in a cubic crystal lattice. The direction of vibrations is $L[100]$

Number of coord. sphere	Body-centred			Face-centred		
	Displaced node	Distance from [[000]]	Number of nodes	Displaced node	Distance from [[000]]	Number of nodes
1	$\frac{1}{2}\frac{1}{2}\frac{1}{2}$	0.867	8	$\frac{1}{2}\frac{1}{2}0$	0.707	8
2	100	1.000	2	100	1.000	2
3	110	1.414	8	$1\frac{1}{2}\frac{1}{2}$	1.225	16
4	$\frac{3}{2}\frac{1}{2}\frac{1}{2}$	1.658	8	110	1.414	8
5	111	1.732	8	$\frac{3}{2}\frac{1}{2}0$	1.581	8
6	200	2.000	2	111	1.732	8

at the origin and interact with it in longitudinal vibrations. The table contains six the nearest neighbours; their coordinates; distance from the origin in periods of the cell and the number of symmetric atoms.

The constructed experimental dispersion curves enable us to make the next step: to the creation of the physical model of interatomic bonds and to computation of the parameters of these models. This problem is difficult and it seems that the measurement of the spectral characteristics of the lattice is the unique direct method, which allows us to half open the curtain over the elaborate laws of mechanisms and phenomena creating the solids as we know them.

Consider the notations of the atomic force constants in the cubic body-centred crystal lattice (Fig. 5.4). The atom located at the origin is affected by a force if another atom in its surrounding is displaced from its equilibrium position. We remember that the atomic force constant is the value of this force per unit of the displacement length. Hence, in the notation of the constant, three parameters must necessarily be reflected:

   a) the direction of the force action onto the chosen initial atom (the first subscript);
   b) the direction of the displacement (the second subscript);
   c) the coordinate of the atom, which is displaced (the Miller indices in double brackets).

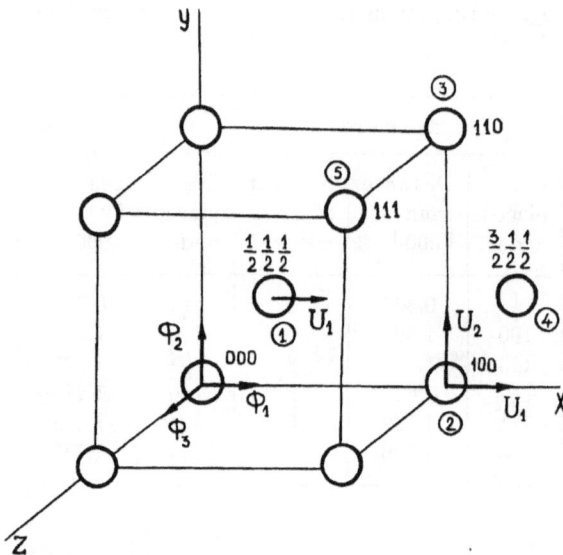

FIGURE 5.4. Scheme of the atoms location in the elementary cell of the cubic body-centred crystal lattice that illustrates the notations of the force constants. The digits in the small circles correspond to ordering of distances from the zero atom located at the origin; some vectors of displacements are shown; the results of these displacements are the forces acting on the initial atom [[000]].

For instance, the constant denoted $\Phi_{11}[[100]]$ is the force arising along the axis Ox if the atom [[100]] is displaced by the length unit in the direction of Ox. Similarly, $\Phi_{12}[[\frac{1}{2}\frac{1}{2}\frac{1}{2}]]$ denotes the force along the axis Ox acting on the atom located at the origin if the atom located at the centre of the elementary cell is displaced along the axis Oy. Other shorter designations of the atomic force constants are also used (see Table 2).

Table 2 contains the results of determination of the values of the atomic force constants for sodium, niobium, and tantalum [15, 64, 65]. This table shows that interatomic interaction in sodium is fulfilled at quite small distances. So, in the third coordination sphere, the force constants are 5–7% of the constants characterizing the interaction of the nearest neighbours. For niobium and tantalum, the existence of long-range interatomic forces is typical. So, even the constants

TABLE 2. Atomic force constants for three metals

Notation		Constants, N/m		
Detailed	Short	Na	Nb	Ta
$\Phi_{11}[[\frac{1}{2}\frac{1}{2}\frac{1}{2}]]$	α_1	1.18	14.14	16.98
$\Phi_{12}[[\frac{1}{2}\frac{1}{2}\frac{1}{2}]]$	β_1	1.32	8.84	11.20
$\Phi_{11}[[100]]$	α_2	0.43	14.20	1.18
$\Phi_{22}[[100]]$	β_2	0.12	−3.64	1.42
$\Phi_{11}[[110]]$	α_3	−0.05	2.27	3.55
$\Phi_{33}[[100]]$	β_3	−0.02	−6.38	−5.43
$\Phi_{12}[[110]]$	γ_2	−0.07	0.76	1.94
$\Phi_{11}[[\frac{3}{2}\frac{1}{2}\frac{1}{2}]]$	α_4	0.04	3.61	3.58
$\Phi_{22}[[\frac{3}{2}\frac{1}{2}\frac{1}{2}]]$	β_4	−0.001	−0.75	−0.72
$\Phi_{23}[[\frac{3}{2}\frac{1}{2}\frac{1}{2}]]$	γ_4	0.005	−0.95	−1.72
$\Phi_{12}[[\frac{3}{2}\frac{1}{2}\frac{1}{2}]]$	δ_4	0.016	1.26	0.98
$\Phi_{11}[[111]]$	α_5	0.017	−1.16	−0.49
$\Phi_{12}[[111]]$	β_5	0.017	−1.33	0.81
$\Phi_{11}[[200]]$	α_6	—	−7.08	−3.71
$\Phi_{12}[[200]]$	β_6	—	1.32	0.13

caused by the affect of eight neighbours are fairly large. Authors of [64] assume that this is connected with the noticeable interaction of quanta of normal vibrations (phonons) with electrons.

5.2. The Fourier analysis of dispersion curves. Interplane force constants

In most cases, researchers use the Born and Karman theory of inter-atomic interaction [6] in order to theoretically process the results of experiments.

Actually, the Born–Karman model does not consider the physical nature of the interaction forces between the atoms in solids. It just assumes that these forces exist and can reveal themselves in attraction of the atoms as well as in their repulsion. The main assumptions of the considered theory can be reduced to three suppositions:

a) the given atom interacts pairwise with one of the neighbouring atoms;
b) the restoring force affecting the given atom linearly depends
on the relative displacement of the other atom;
c) in the expansion of the potential energy of the interaction in
powers of atomic displacements, one can neglect the terms
containing the powers of displacements higher than second.

It is necessary to take into account that, in the Born–Karman
model, for lattice vibrations, the forces affecting the given atoms by
displaced atoms are summed.

The main question of the Born–Karman theory is the following:
what is the number of the nearest neighbours, which interact with
the atom chosen as the origin. In other words, by how many nearest
neighbours we can restrict ourselves when studying the interatomic
interactions?

Foreman and Lomer were the first researchers to report the fact
that, if the vibrations propagate in the cubic lattice along the high
symmetrical directions <100>, <110>, <111>, then the secular equation (1.52) is simplified and can be solved directly [66]. This corresponds to the following physical treatment. The lattice waves propagating along the symmetrical directions correspond to the vibrations
of parallel atomic planes as whole objects so that the mathematical
description of the process can be reduced to a linear chain. Every
atom in this chain interacts with the neighbours in harmonic approximation and every chain link represents an atomic plane. In this
case, in the face-centred lattice, for instance, one longitudinal and
two transverse modes are independent; thus, every oscillatory mode
can be treated as related to a separate linear chain.

If this is true, then the cyclic frequency of the atomic plane vibrations ω must satisfy the equation similar to (1.20):

$$m\omega^2 = \sum_{p=1}^{N} F_p \left(1 - \cos pqb\right), \qquad (5.1)$$

where the coefficients F_p can be called the interplane force constants;
b is the shortest distance between the parallel atomic planes. Thus,
the interplane force constants for the directions <100>, <110>, <111>
are the Fourier coefficients in the expansion of the squared frequencies
of vibrations as a function of the wave vector.

The quantity N is the number of atomic planes, the interaction with which is essential; the interaction with planes which are farther from the atom can be neglected, thus, $F_p = 0$ for $p > N$.

This implies an effective rule of estimating the distances at which the interatomic forces act in one or another solid body. The number itself of the terms in the Fourier expansion of the experimentally obtained function $\omega^2(q)$ is indicative of the limit of action of the interatomic forces. The number of the last coordination sphere, where the coefficient F_p is essential, corresponds to the boundary of the interaction between the atoms.

If the number of the terms of the series or, which is the same, the limit of action of the interatomic forces is not very large, then the atomic force constants can be calculated on the basis of data for the determination of F_p. The point is that the interplane force constants represent a linear combination of the atomic force constants. The concrete equation for determination of the atomic constants depends on the type of the crystal lattice and on the oscillatory mode.

In the first section, we have considered the expansion of the function $\omega^2(q)$ into the Fourier series. There, we have also presented appropriate formulae (1.20)–(1.24). It is expedient to change slightly the work formulae for the calculations by using the following relations:

$$q_m = \frac{2\pi}{2b}; \qquad (5.2)$$

$$\zeta = \frac{q}{q_m}; \qquad (5.3)$$

$$q = \frac{\zeta\pi}{b}, \qquad (5.4)$$

where b is the smallest distance between the projections of the atoms onto the direction of the vibrations. For instance, for the cubic lattice, the distances b are equal to $a/2$, $a\sqrt{2}/2$, and $a\sqrt{3}/6$ in the directions [100], [110], and [111], respectively. Here, a is the period of the elementary cell.

Substituting (5.2)–(5.4) into formula (1.20), we obtain the following expression for the expansion of the squared angular frequency into the Fourier series:

$$\omega^2(q/q_m) = \frac{1}{m}\left\{ F_0 - \sum_{p=1}^{N} F_p \cos\left(\frac{\pi pq}{q_m}\right)\right\}, \qquad (5.5)$$

$F_0 = \sum_{p=1}^{N} F_p$; p is the number of the coefficient; N is, as before, the number of planes, the interaction with which is essential for the given atom. The parentheses on the left-hand side of the formula mean the functional dependence.

The quantities F_1, F_2,..., F_N represent the interplane force constants. They have a clear physical sense. The coefficient F_1 is equal to the force, which acts on the chosen atom if all atoms in the two nearest neighbouring planes located on each side of the given plane are displaced at the unit distance along the normal to this plane. Similarly, the coefficient F_2 is determined by the force appearing while displacing the second atomic plane at the unit distance etc. The displacement of the planes can be longitudinal for longitudinal vibrations and transverse for transverse ones.

The interplane constants are determined by the following formulae:

$$F_0 = m \int_0^1 \omega^2(\zeta)\, d\zeta; \qquad (5.6)$$

$$F_p = -2m \int_0^1 \omega^2(\zeta)\, \cos(\pi p \zeta)\, d\zeta. \qquad (5.7)$$

The data of computations of the interplane force constants and group velocities of elastic waves of the lattice are presented for several metals. As the initial data for the computations, we used the arrays of experimental dependences of the squared frequencies of the atomic vibrations on the ratio q/q_m. At the first stage, the number of interacting atomic planes must be chosen. In other words, first, the value of N in formula (5.5) should be found. The idea was to compare the experimental values of the dependence $\nu(q)$ with those calculated by equation (5.5) obtained with the use of a series of the interplane coefficients. This comparison was made consecutively by varying the number N of planes taken into account from 2 to 19. As the criterion of consistency of the experimental values with the theoretical ones, the least value of the sum of the mean-square deviations of the calculated values of the frequency from the experimental ones was chosen:

$$\Delta\nu^2 = \sum_{q/q_m=0}^{1} (\nu_e - \nu_c)^2. \qquad (5.8)$$

The sum is taken over all values of q/q_m from 0 to 1 with the step 0.1. On the base of the obtained data, the problem of determination of the minimal number of interacting atomic planes is solved. usually, the function $\Delta\nu^2$ depending on N passes through the minimal value; oscillations of this quantity are also observed.

After the number N was found, the values of interplane force constants were calculated by formulae (5.7).

Consider the results of the Fourier analysis of the dispersion curves for a number metals: sodium, niobium, molybdenum, tantalum, tungsten and lead. The experimental measurements allowing to construct the curves $\omega^2(\zeta)$ were fulfilled by the authors of papers [14, 15, 64, 65, 67, 69].

Table 3 contains the results of calculations of the interplane force constants for the mentioned metals. In the same table, the values of the number of atomic planes are also given for which the sum of the deviations $\Delta\nu^2$ according to formula (5.8) is minimal.

In Fig. 5.5, the dependences of these sums on the number of terms in the expansion for sodium and niobium are given. In the body-centred cubic crystal lattice, the least distance between the neighbouring atoms is observed in direction [111]. It is natural to assume that, for the case of longitudinal vibrations, the interatomic interaction must also be the greatest in this direction. As the number of accounted neighbour interactions grows, the value $\Delta\nu^2$ oscillates, showing a deep relative minimum at $N = 7$ ($\Delta\nu^2 = 0.02$). Hence, in the crystal lattice of sodium, when the longitudinal wave propagates along [111], i.e., along the solid diagonal of the elementary cell, the interaction between seven nearest atomic planes of the type (111) is essential. For niobium, the curve also oscillates; however, it tends to the absolute minimum ($\Delta\nu^2 = 0.15$) only at $N = 15$. Hence, in the framework of the Born–Karman model, the interatomic interaction in the crystal lattice of niobium in the given direction extends much farther than in the crystal lattice of sodium.

However, in sodium, the transverse vibrations $T[111]$ cause the stack of atomic planes to be twice as "thick", namely, 15. It should be noted that the relative minima on the curve $\Delta\nu^2(N)$ are observed at $N = 4, 7, 10$. Maybe, this phenomenon should be treated as essential interactions at small distances and weak, gradually attenuated at

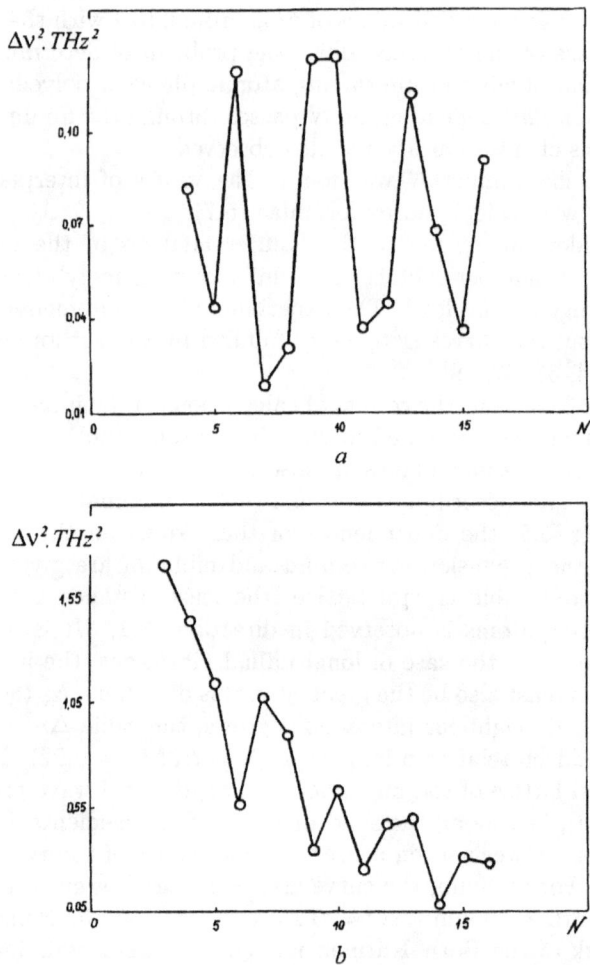

FIGURE 5.5. The influence of accounting of the number of atomic planes interacting with each other on the mean-square deviation of the theoretical values of the frequency from the experimental ones:
a — sodium, the oscillatory mode $L[\zeta\zeta\zeta]$; b — niobuim, $L[\zeta\zeta\zeta]$; c — sodium, $T[\zeta\zeta\zeta]$; d — niobium, $T[\zeta\zeta\zeta]$. (Comparing the graphs with each other, one should take into account the difference of scales along the y-axis.)

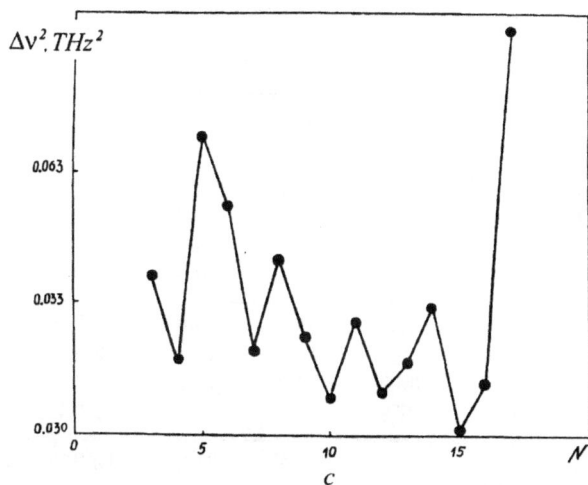

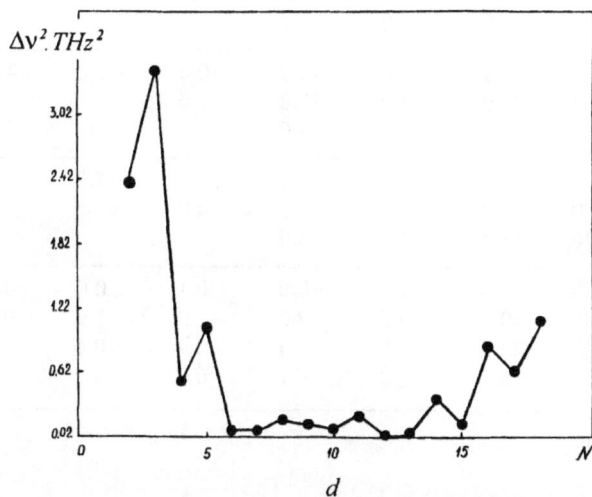

FIGURE 5.5 (*continued*).

large distances. In niobium, the vibrations $T[111]$ spread over 6–7 interatomic distances and strongly attenuate up to 12 neighbour planes.

The values of the interplane force constants in sodium are by an order of magnitude, less than in niobium. In the case of vibrations

TABLE 3. Interplane force constants (N/m) of metals in different crystallographic directions.

Metal	[uvw]	F_1	F_2	F_3	F_4	F_5	F_6	N
	$L[\zeta\zeta\zeta]$	1.65	1.50	7.75	−0.36	0.14	0.11	7
Na	$T[\zeta\zeta\zeta]$	9.85	1.53	−0.27	−0.03	0.07	−0.16	15
	$L[\zeta00]$	9.34	1.00	0.38	−0.02	−0.10	0.11	15
	$T[\zeta00]$	9.84	0.18	−0.10	−0.03	−0.06	0.05	13
	$L[\zeta\zeta\zeta]$	36.8	17.1	81.6	−9.0	9.7	−7.0	14
Nb	$T[\zeta\zeta\zeta]$	113.0	9.3	16.4	−14.1	−2.3	4.4	7
	$L[\zeta00]$	101.9	36.3	24.2	−14.5	2.8	2.7	8
	$T[\zeta00]$	135.3	−27.9	−5.9	4.6	−1.7	1.6	9
	$T[\zeta\zeta\zeta]$	113.0	86.0	−0.2	21.0	−12.0	−1.0	
Mo	$L[\zeta00]$	122.0	116.0	−12.0	17.0			
	$T_2[\zeta\zeta0]$	77.0	4.0	7.0				
	$T[\zeta\zeta\zeta]$	133.0	5.0	17.0	−10.0	−1.0	4.0	
Ta	$L[\zeta00]$	131.0	27.0	21.0	−7.0			
	$T_2[\zeta\zeta0]$	109.0	−21.0	4.0				
	$T[\zeta\zeta\zeta]$	183.0	97.0	0.1	14.0	−4.0	−6.0	
W	$L[\zeta00]$	190.0	120.0	−7.0	4.0			
	$T_2[\zeta\zeta0]$	115.0	10.0	−5.0				
	$T[\zeta00]$	6.6	4.5	−0.9	0.1	0.0	−0.1	
Pb	$L[\zeta00]$	29.5	9.7	−2.6	−1.9	−1.9	0.9	
	$L[\zeta\zeta\zeta]$	29.6	5.8	1.4	0.4	0.4	0.1	
	$T[\zeta\zeta\zeta]$	5.2	0.3	0.1	0.2	0.2	−0.1	

in the body-centred lattice along [111], the coefficients F_p are peculiarly ordered: the greatest is the third one, followed by the first and the second. For the transverse vibrations, the first coefficient is the largest. As a rule, the fourth one is negative. The absolute value of far coefficients can be quite large. For instance, in niobium, the seventeenth and nineteenth coefficients are negative and comparable with the first three coefficients. It does not seem that in niobium the value of the interplane force constants becomes, starting from some value of N, negligibly small.

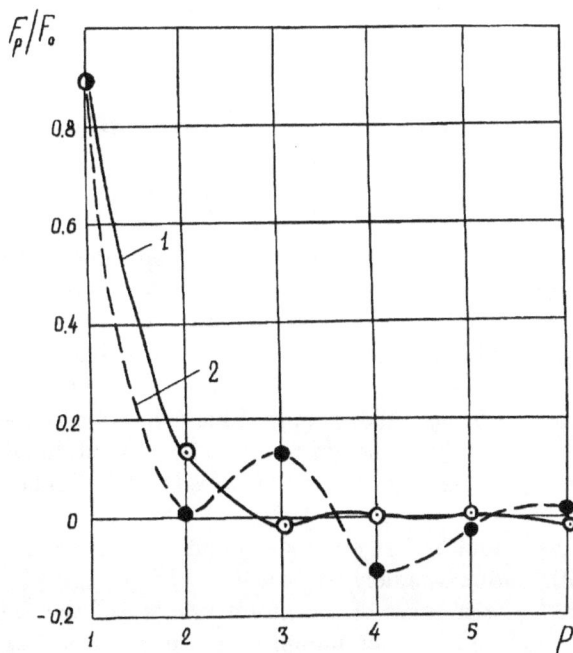

FIGURE 5.6. The dependence of the ratio of the constants F_p/F_0 on the number of the interplane force constant for the mode $T[\zeta\zeta\zeta]$:
1 — sodium; 2 — niobium.

In Fig. 5.6, the dependences of the ratio F_p/F_0 on the number of the coefficient are given for the same two metals. Considering the ratio, we can exclude the differences in absolute values of the constants. The curves go similarly; however, for niobium, essential oscillations are typical. The figure shows that the interatomic interaction in sodium attenuates in this crystallographic direction at much smaller distances than in niobium.

The obtained coefficients F_p enable us to calculate the group velocity of the elastic wave in the chosen crystallographic direction, as well as the dependence of this velocity on the wave vector. The group velocity of the wave v_g is the derivative of the cyclic frequency with respect to the wave vector:

$$v_g = \frac{d\omega}{dq}. \tag{5.9}$$

Differentiating expression (5.5) and taking into account (5.3), we obtain

$$v_g = \frac{b \sum_{p=1}^{N} p\, F_p \, \sin(\pi p \zeta)}{2m\omega}. \tag{5.10}$$

We now review the data concerning the dependence of the group velocity of vibrations in the lattice on the wave vector. The curves obtained for the same two metals (Na, Nb) can be seen in Fig. 5.7. The abscissa of the dependence $v_g(q/q_m)$ is divided in three equal parts: at the corresponding points, the wave velocity is equal to zero. These points are $0.33q_m$; $0.67q_m$; q_m. This experimental fact implies the following conclusion: the group velocity of longitudinal waves in the direction [111] vanishes when the wave length is equal to $6b$, $3b$, or $2b$.

Note that niobium has one more point of zero velocity in comparison with sodium, namely, at $\lambda = 4b$.

In Fig. 5.8, the curve of the group velocity of the longitudinal oscillatory mode [110] in aluminium is shown.

Figure 5.9 illustrates the scheme of displacements of a family of parallel atomic planes when a system of standing waves is excited in this family. The displacements of each plane under the vibrations are shown by arrows. One can, for instance, see that, if $q/q_m = 1$ and $\lambda = 2b$, then the neighbouring planes are displacing toward each other remaining in antiphase. This is the shortest wave length which can propagate in the crystal lattice.

If $q/q_m = 0.67$, $\lambda = 3b$, then planes 1, 4, 7,... are at rest; planes 2 and 3, 5 and 6, ... are displacing in opposite directions. The further, for the cases $\lambda = 4b$, $5b$, $6b$ is clear from the figure.

It is interesting that neither in sodium, nor in niobium, and in general, in any of the studied metals, the case of the standing wave such that $\lambda = 5b$ and $q/q_m = 0.4$ is realized. In this case, planes 1 and 6 should be immobile; planes 2 and 3, 4 and 5 should be displacing in antiphase; this is not observed. The case $q/q_m = 0.5$, where the alternate planes vibrate in antiphase, is typical for niobium and is not observed in sodium (Table 4).

Positions of extrema on the dispersion curves for different metals are presented in Table 4.

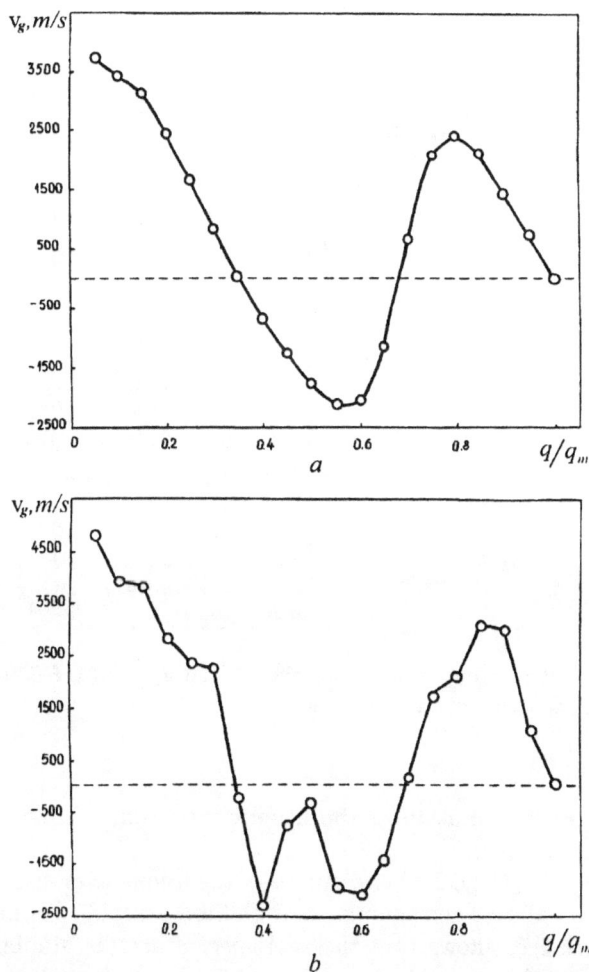

FIGURE 5.7. The group velocity of the mode $L[\zeta\zeta\zeta]$ as a function of the wave vector:

a — sodium; b — niobium.

5.3. Frequency spectrum of real metals

On the basis of dispersion curves obtained directly by the method of nonelastic scattering of neutrons, many authors constructed the

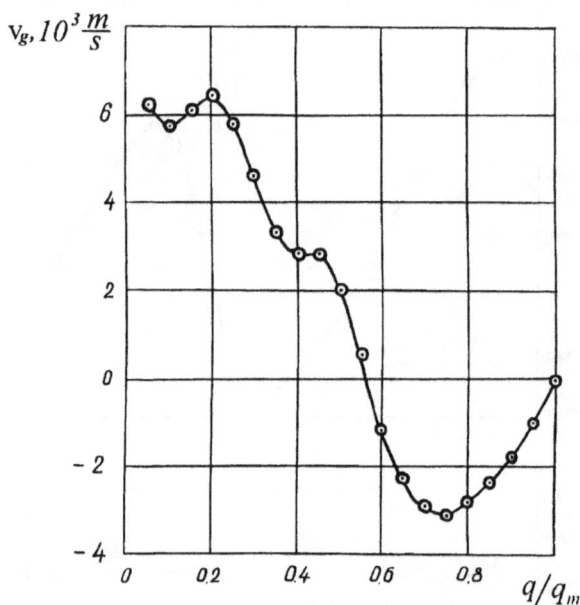

FIGURE 5.8. The group velocity of the oscillatory mode $L[\zeta\zeta 0]$ in aluminium as a function of the wave vector.

curves of the distribution of the atomic vibrations with respect to the frequencies $g(\nu)$.

In Fig. 5.10, the calculated curves of the frequencies distributions in diamond, silicon, germanium, and gallium arsenide are presented [70]. Figure 5.10 shows that the real spectra are not similar to the Debye model. A representative high peak is observed on all the curves in the high-frequency domain. It is somewhat wider for diamond and is double for the other materials. The curves for diamond are different from the other three. Obviously, the curve for germanium is similar to that for the semiconductor GaAs. The direct proportional dependence is seen between the frequency corresponding to this peak for different elements of the IV group and the values of $\sqrt{\alpha/m}$, where α is the force constant and m is the mass of the atom. In Fig. 5.11, we have $\omega = 4\sqrt{\alpha/m}$. In other words, the relation typical for simple spring pendulum holds.

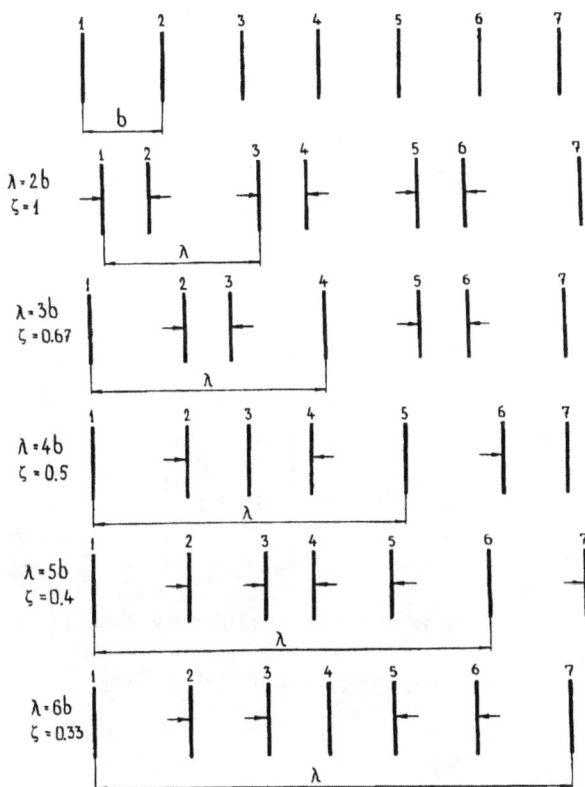

FIGURE 5.9. The scheme of displacements of the atomic planes under longitudinal vibrations in the crystal lattice. The first row corresponds to unperturbed lattice. The next rows illustrate the formation of a system of standing waves. The arrows show the vectors of longitudinal displacements.

The curves of spectral density for aluminium [71] and γ-iron [72] are typical for metals, Fig. 5.12. The structure of the curves is similar: relatively diffuse low-frequency part and high, narrow high-frequency peak. Its position is analogous to the frequency of the optical modes in the twoatomic lattice and is caused, obviously, by displacements of the atomic planes toward each other.

The method of nonelastic scattering of neutrons indicates that the range of interatomic interaction is different for different elements.

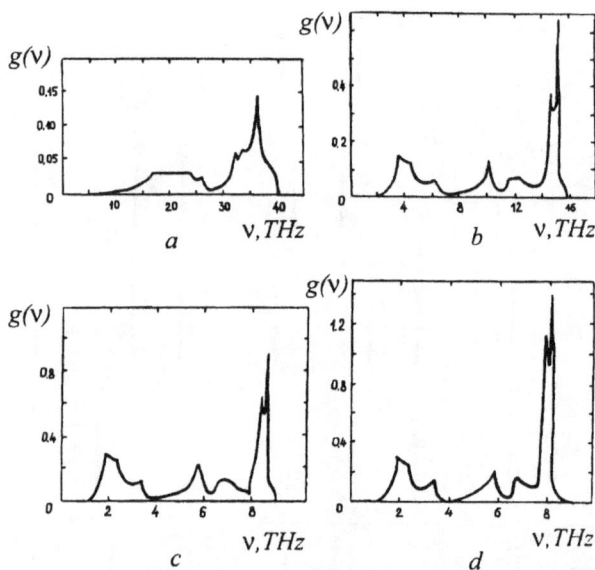

FIGURE 5.10. The curves of the spectral density distribution of atomic vibrations:

a — diamond; b — silicon; c — germanium; d — GaAs.

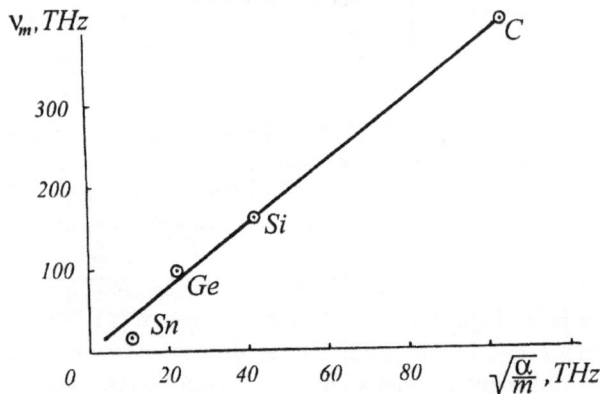

FIGURE 5.11. The dependence of the frequency corresponding to the high-frequency peak on the curves $g(\nu)$ on the values of $\sqrt{\alpha/m}$ for the four materials.

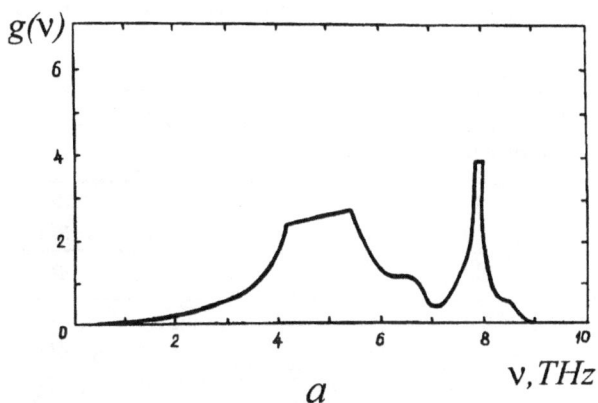

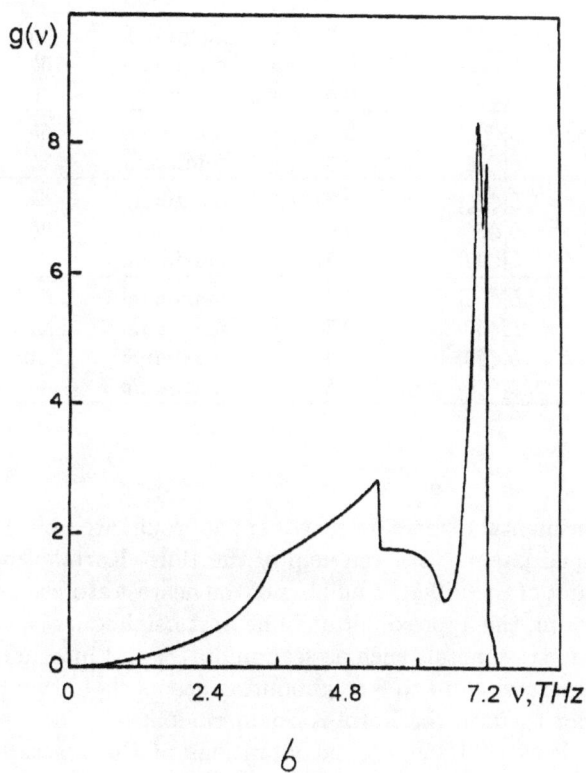

FIGURE 5.12. The distribution of the spectral density of atomic vibrations:
a — aluminium; b — γ-iron at the temperature 1428 K.

TABLE 4. Positions of extrema on the dispersion curves and lengths of standing waves

Metal	Direction of vibrations	Position of extremum, ζ	Form	λ	Source
Na	$L[\zeta\zeta\zeta]$	0.33	maximum	$6b$	[15]
	$L[\zeta\zeta\zeta]$	0.66	minimum	$3b$	
Al	$L[\zeta\zeta 0]$	0.5	maximum	$4b$	[63]
Nb	$L[\zeta 00]$	0.5	maximum	$4b$	[64]
	$L[\zeta\zeta 0]$	0.5	maximum	$4b$	
	$L[\zeta\zeta\zeta]$	0.33	maximum	$6b$	
	$L[\zeta\zeta\zeta]$	0.69	minimum	$3b$	
Mo	$L[\zeta 00]$	0.66	maximum	$3b$	[69]
	$L[\zeta\zeta\zeta]$	0.33	maximum	$6b$	
	$L[\zeta\zeta\zeta]$	0.67	maximum	$3b$	
$Ta_{77}W_{23}$	$L[\zeta\zeta\zeta]$	0.33	maximum	$6b$	[68]
	$L[\zeta\zeta\zeta]$	0.67	minimum	$3b$	
$Ta_{33}W_{67}$	$L[\zeta\zeta\zeta]$	0.33	maximum	$6b$	[68]
	$L[\zeta\zeta\zeta]$	0.67	minimum	$3b$	
	$L[\zeta 00]$	0.67	maximum	$3b$	
Pb	$L[\zeta 00]$	0.75	maximum	$2.67b$	[14]
	$T[\zeta 00]$	0.60	maximum	$3.33b$	
	$L[\zeta\zeta 0]$	0.84	maximum	$2.36b$	
	$L[\zeta\zeta\zeta]$	0.5	maximum	$4b$	

The experimental dispersion curves are approximated by the authors of published papers with the help of the Born–Karman model with due account of the different number of the nearest atoms. For iridium and rhodium, the approximation of nearest neighbours is well consistent. As for the metals such as scandium [78] and lutecium [79], the mutual interaction up to 8 neighbouring atoms should be taken into account for them in the Born–Karman model.

The density of the normal vibrations of the crystal lattice of vanadium at the room temperature has been measured [92]. The curve is characterized by the presence of the peaks at the frequencies 4.9 and 6.9 THz; the authors relate these peaks with longitudinal and transverse vibrations, respectively. The dependence $g(\nu)$ drops abruptly to zero at 8.1 THz.

The similar spectra are constructed for many elements and chemical compounds.

5.4. Influence of ordering in intermetallides

The curves of density of the vibration states are ordering-sensitive. The curves $g(\nu)$ for the intermetallide Ni_3Al have been measured in the ordered and nonordered states [73]. It was found that in the ordered state (atoms of aluminium are at vertices of the cubic face-centred elementary cell, atoms of nickel are at the centres of the faces), a high-frequency peak arises in the domain of the optical mode. The difference of the curves in frequency reaches 9.41 THz. This difference is connected with the vibrations of the aluminium sublattice.

The influence of ordering was also studied for the intermetallides Cu_3Au [74, 75] and FeAl [76]. The extension of the oscillatory modes under ordering, essential anisotropy and anharmonicity of the atomic vibrations are noted. For instance, in Cu_3Au, the atom [[1/2 1/2 0]] vibrates, mainly, in the direction [001]. In general, the ordering process leads to the change of the force constants of the interatomic interactions. In FeAl, the energy of the pairwise interaction of atoms Fe–Al is noticeably higher than those of atoms Fe–Fe.

It is interesting to consider the influence of the chemical composition of solid solutions on the example of compounds niobium–molybdenum [80]. Both the metals belong to transition metals with the unfilled electron $4d$-shell. Pentavalent niobium has five electrons and hexavalent molybdenum has six $(d + s)$ electrons. Figure 5.13 shows the dispersion curves for niobium, molybdenum, and two compounds of intermediate composition. In molybdenum, the frequencies of the oscillatory modes are much greater. Since the difference of the atomic masses is rather small ($M_{Mo} = 95.94$, $M_{Nb} = 92.91$), the matter is, obviously, associated with the force constants, which are greater for molybdenum. The intersection of L- and T-modes disappears as the concentration of molybdenum increases. The interatomic forces are very long-ranged: according to the data of the analysis of the Fourier coefficients for separate oscillatory modes, these forces range up to the 13th neighbour.

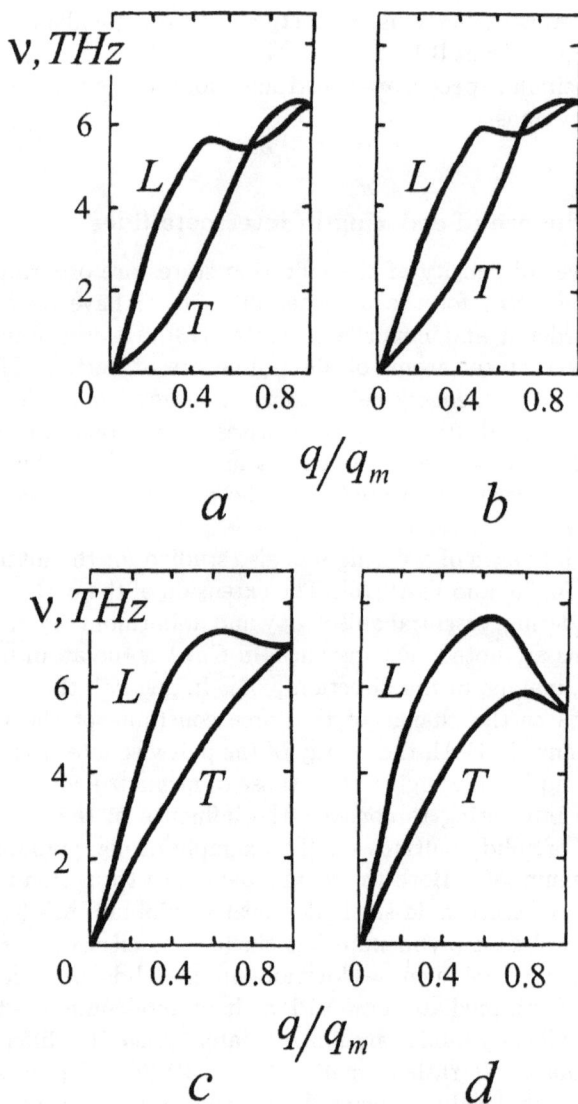

FIGURE 5.13. The dispersion curves for compounds of the system Nb–Mo:
a — pure niobium; b — $Nb_{0.85}Mo_{0.15}$; c — $Nb_{0.25}Mo_{0.75}$; d — pure molybdenum.

The simple, logical and evident Born–Karman model is the model of "rigid ions". In many cases, it does not give satisfactory results even if large number of interactions is taken into account. In order to make the calculated and experimental dispersion curves consistent, the so-called fitting parameters should be introduced. The number of these parameters is up to 15–20. Unfortunately, many of them have no definite physical sense. The detailed description of numerous proposed models is beyond the scope of this book.

Very thorough and extensive study of the dispersion of the atomic vibrations in the compound GaAs has been reported [81]. The dispersion was studied at the temperature 12 K in six crystallographic directions. In the study, eight models of interatomic interactions were analyzed successively and quantitatively. As an example, we briefly mention them: the model of rigid ions; shell models; the model of deformation dipoles; the model of overlapping electron shells; the model of charged bond.

The model of rigid ions contains constants corresponding to the nearest neighbours, the second neighbours, and the Coulomb forces between atoms. Several authors assumed that the potential energy of the vibrating atom depends not only on the atoms displacements, but also on the dipole moments induced as a result of vibrations. With due account of this fact, the shell models assume that ions are deformable mechanically and electrically. These models take into account the polarizability and contain a large number of parameters. The model of overlapping valent shells include into itself the parameters of deformation of the shells. In the model of charged bond, ions are assumed nonpolarizable; the Coulomb forces and forces depending on the angle act. Just for this model, the best agreement of the experimental and calculated data was obtained in the study described [81].

In general, the results of discussions on the acceptability of different models of the interatomic bonds on the base of the dispersion curves are of interest, however, cannot be treated as single-valued.

5.5. The rule of sums

Thus, the assumptions on the nature of the interatomic forces imply a unique calculated totality of $3n$ branches of frequency ω of the ith oscillatory mode, which depends on the wave vector q. Here, $i = 1$, $2, \ldots, 3n$, where n is the number of atoms in the elementary cell.

The inverse transition from the curves $\omega(q)$ to the set of constants is not unique, especially, taking into account the limited number of observations and the experimental errors. Actually, the agreement of the theory and experiments can be achieved for any force schemes and a sufficient number of parameters.

This is the reason, why it has been proposed to replace the analysis of dispersion curves by the study of the so-called "sum function" $\sum_{i=1}^{3n} \omega_i^2(q)$. Obviously, the sum of the squared cyclic frequencies do not contain the complete information, however, it enables one to obtain some essential results in a simpler way than the analysis of the dispersion curves.

For the first time, the rule of sums has been proposed by Brout [82] for ionic crystals of the type NaCl, where two kinds of forces act between the ions: the Coulomb attraction forces and the repulsion forces. For the crystal lattice of tetrahedral symmetry with two unpolarizable ions in the elementary cell, we have

$$\sum_{i=1}^{6} \omega_i^2(q) = \frac{18r_0}{\beta} \left(\frac{1}{m_+} + \frac{1}{m_-} \right), \tag{5.11}$$

where ω_i it the cyclic frequency of the ith branch of the dispersion curve for given q; r_0 is the distance between the ions; β is the compressibility coefficient; m_+ and m_- are the masses of positive and negative ions, respectively. Brout proved equation (5.11) by analyzing the trace of the dynamic matrix. Later, Rosenstock showed [83, 84] that the equation

$$\sum_{i=1}^{3n} \omega_i^2(q) = \text{const} \tag{5.12}$$

can be employed in order to extract the information on the interatomic forces acting in nonionic crystals. Equation (5.12) shows that its right-hand side does not depend on q. It is the author's opinion that this equation holds for crystals which belong to one of the two classes: for the first one, forces of electromagnetic nature (Coulomb forces, dipoles, multipoles) are typical; for the second class, the interactions between atoms of different kind are typical. In materials belonging to these classes, Rosenstock called the interatomic forces "trace-constant". This rather exotic name is due to the behaviour of

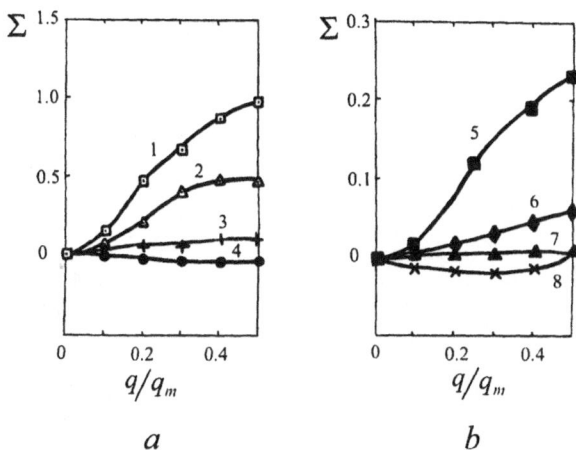

FIGURE 5.14. The sums of the squared frequencies of the oscillatory modes as functions of ζ:
1 — molybdenum; 2 —tungsten; 3 — tantalum; 4 — niobium; 5 — diamond; 6 — GaAs; 7 — silicon; 8 — germanium.

the trace of the dynamic matrix when the wave vector varies. Accordingly, the forces between atoms in compounds belonging to neither of these classes were called by the same author "trace-variable".

The calculated data of sums of the squared frequencies for a number of elements [84] are shown in Fig. 5.14; tantalum and niobium differ from molybdenum and tungsten. The force interaction in the first pair has, in Rosenstock's terms, the eletromagnetic nature. Diamond remains different from the semiconductors; the constancy of the sum is not observed in it, contrary to silicon and germanium. The authors believe that this simple method enables the separation of the long-range electromagnetic forces from the forces caused by overlapping of the electron shells. Among the first are the following elements and compounds: Mo, W, β-Sn, Pb, diamond, GaAs, Zn. The second group consists of Na, Ta, Nb, Si, Ge, Ni, Be. Obviously, this classification should be used with care. For instance, it is known that diamond is a classical example of coalescence of the wave functions of neighbouring atoms and formation of covalent bonds.

6. ANHARMONICITY OF ATOMIC VIBRATIONS

We consider the process of atomic vibrations of a crystal lattice, and three possible approximations can be made.

The first approximation is called harmonic. It is characterized by the fact that, in the expansion of the potential energy of interatomic interaction (1.7) and (1.41), we neglect all the terms except the quadratic one. This means that the curve of the dependence of the potential energy on the interatomic distance is approximated at the minimum by a parabola. As a result, if the distance between the atoms increase or decrease, the interaction energy is changed by the same value. The harmonic approximation allows us to simplify the description of the mechanism of phenomena; however, it implies a number of suppositions, which are not confirmed by experiments. For example, the harmonic approximation implies that there are no thermal expansions of solid bodies, the force constants are independent on the temperature, the coefficient of thermal conductivity is infinite, there are no interactions between the vibrations. All of these conclusions contradict the experimental facts.

The quasiharmonic approximation is based on the assumption that solid bodies expand when heating and only this expansion affects the force constants and vibration frequencies. The increase of the interatomic distances results in a decrease ("softening") of the force constants and in diminishing of the frequency of normal oscillatory modes. In the quasiharmonic theory, the concept of the Gruneisen constant γ_G as a measure of deviation of the crystal from the harmonicity is introduced.

At last, the anharmonic approximation takes into account terms of the series expansion of the potential energy whose degree is higher than two.

133

6.1. The general theory of accounting the terms of higher order

We return to the consideration of the one-dimensional chain consisting of atoms of the same kind. Suppose that the neighbouring atoms are equidistant, each atom interacts with all the others and one can neglect the boundary effects. The initial atom is given number zero; the next, number one, etc. The number of the last atom is n, and there are totally $n + 1$ atoms. And, n is assumed sufficiently large so that the product $na \gg a$. The distance from the origin is $r = na$.

We recall that the first term of the expansion of the potential energy $\Phi(r)$ into the Taylor series just shifts the origin at another point , and the second term is equal to zero. The complete potential energy of interaction can be written as

$$\Phi(r) = \sum_{i=0}^{n-1} \sum_{p=1}^{n-i} \left[\frac{1}{2} \left(\frac{\partial^2 \Phi}{\partial r^2} \right)_{r=pa} (u_{i+p} - u_i)^2 \right.$$
$$\left. + \frac{1}{6} \left(\frac{\partial^3 \Phi}{\partial r^3} \right)_{r=pa} (u_{i+p} - u_i)^3 \right]. \quad (6.1)$$

Here, i is the number of an atom and p is a positive integer 1, 2, 3, 4,...; thus, $i + p$ denotes the number of the atom which interacts with the atom number i. Then, we must use the Newton second law and write the motion equation in the form

$$-\frac{\partial \Phi}{\partial r} = m \frac{\partial^2 u}{\partial t^2}. \quad (6.2)$$

The solution of this equation is very difficult, especially, in the case, where the anharmonicity is noticeable. This is due to the propagation of the perturbations in a nonlinear system; contrary to the harmonic chain, the force affecting the atoms is not simply proportional to the displacement.

Therefore, it is expedient to consider a simpler model [5]. The case at hand is a twoatomic molecule. It can be described by the Morse potential:

$$\Phi(r) = D\{1 - \exp[-\lambda(r - a_0)]\}^2, \quad (6.3)$$

where D is the dissociation energy; λ is a positive constant; a_0 is the average distance between atoms at $T = 0$. For high temperatures, we have $a > a_0$ and the curve deviates from the parabola (Fig. 6.1).

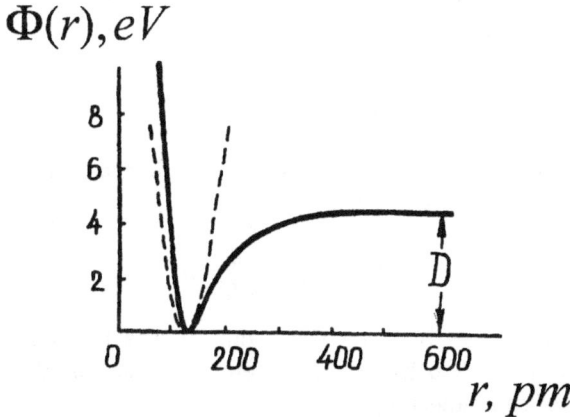

FIGURE 6.1. The dependence of the energy of interatomic interaction on the distance for twoatomic molecule of the type HCl.

Denote the atomic displacement by $u_0 = r - a_0$ and expand the potential energy near the point a_0 into the power series:

$$\Phi(r) = \frac{1}{2}f_0 u_0^2 + \frac{1}{6}g_0 u_0^3 + \dots . \tag{6.4}$$

Here, the force constants f_0 and g_0 denote the values of the second and third derivatives at the point of minimum, respectively.

Using the Morse potential and computing the derivatives, we obtain

$$f_0 = 2\lambda^2 D; \qquad g_0 = -6\lambda^3 D. \tag{6.5}$$

The constant g_0 is negative.

The motion equation can be written in the following form:

$$\frac{\partial^2 u_0}{\partial t^2} + \frac{f_0}{m}u_0 + \frac{g_0}{2m}u_0^2 = 0. \tag{6.6}$$

Then, we denote $f_0/m = \omega_0^2$ and $g_0/2f_0 = s$.

The solution has the form

$$u_0 = v_0 + A(\cos \omega t + \eta \cos 2\omega t). \tag{6.7}$$

This means that the expansion of the potential energy illustrated by the solid curve in Fig. 6.1 contains not only the term corresponding to the principal frequency of vibrations ω but also higher vibration harmonics with frequencies 2ω, 3ω etc.

Substituting (6.7) into the main equation (6.6) and neglecting some values, we can obtain

$$a = a_0 - \frac{1}{4}\frac{g_0}{f_0}A^2, \qquad (6.8)$$

i.e., $a > a_0$ since $g_0 < 0$. The period is varying, as the temperature increases, by the following rule:

$$a(T) = a_0 - \frac{1}{2}\frac{g_0}{f_0^2}k_BT. \qquad (6.9)$$

Thus, the distances between the atoms in the molecule grows directly proportional to the absolute temperature. For the vibrations frequency of the molecule, we have

$$\omega^2(T) = \omega_0^2\left(1 - \frac{g_0^2}{f_0^3}k_BT\right). \qquad (6.10)$$

From this formula, one can see that the frequency of vibrations of the quasiharmonic crystal decreases: $\omega < \omega_0$.

The harmonic crystal does not increase its size as the temperature increases, because the constant $g_0 = 0$ in this case. The anharmonicity reveals itself in increasing the distance between the atoms; the second evidence is the decrease of the frequency of atomic vibrations. Both these effects are consequences of the diminishing of the values of the force constants of the material. For instance, the constant f_0 is decreasing as the temperature increases in accordance with the following equation [5]:

$$f = f_0 - \frac{1}{2}\frac{g_0^2}{f_0^2}k_BT + \frac{3}{8}\frac{h_0}{f_0}k_BT. \qquad (6.11)$$

6.2. The coefficients of the potential energy expansion

To determine the characteristics of the anharmonicity experimentally, we must concretize the equations for the potential energy expansion. They should be represented in the form, which allows us to use the parameters measured by diffraction methods. As has been said, in the harmonic approximation we have

$$\Phi_k = \Phi_0 + \frac{1}{2}\alpha_k\overline{u_k^2}, \qquad (6.12)$$

where k denotes a direction of displacement; the force parameter α_k is connected with mean-square displacement by the relation

$$\alpha_k = \frac{k_B T}{\overline{u^2}}. \tag{6.13}$$

(The coefficient α_k in formula (6.4) is denoted by f_0.)

With due account of the expansion of the atomic displacements along the axes, we have

$$\overline{u_k^2} = \overline{u_1^2} + \overline{u_2^2} + \overline{u_3^2}. \tag{6.14}$$

The corrections for the anharmonicity in the potential energy of the atoms interaction Φ_k are represented by the terms of third and higher degrees added to the right-hand side of equation (6.12). These corrections correspond either to an isotropic crystal or to an anisotropic one. In the first case, they depend only on the modulus of the displacement. In case of the isotropy, the additional terms have the following form [85]:

$$\gamma_k \left(u_1^2 + u_2^2 + u_3^2 \right)^2 \quad \text{and} \quad \varepsilon_k \left(u_1^2 + u_2^2 + u_3^2 \right)^3. \tag{6.15}$$

The parameters γ_k and ε_k are negative. The terms corresponding to the anisotropic crystal take into account the angular dependence of the potential energy. For instance, for the structure of diamond, the term of the third degree has the form:

$$\beta_k u_1 u_2 u_3. \tag{6.16}$$

The concrete expression for the potential energy depends on the symmetry of the crystal. Thus, the experimental study of the anharmonicity is reduced to the determination of the values of coefficients of the types α, β, γ, ε, and δ.

6.3. The influence of the anharmonicity on the frequency spectrum of vibrations

For relatively high temperatures, the contribution of the lattice vibrations in the value of the free energy is [4]

$$\Delta F = k_B T \sum \ln \left[\frac{h \omega_j(\mathbf{q})}{2\pi k_B T} \right], \tag{6.17}$$

where, as usual, j is the number of the oscillatory mode, whose frequency depends on the wave vector. If the cubic crystal extends from a volume V to $V + \Delta V$, then its energy increases by the quantity

$$\Delta U = \frac{\Delta V^2}{2\varkappa V},$$ (6.18)

where $\varkappa$ is the coefficient of compressibility: $\varkappa = \dfrac{\Delta V/\Delta p}{V}$. Therefore, the free energy can be expressed as

$$F = \Delta F + \Delta U.$$ (6.19)

The equilibrium condition has the following form:

$$\left(\frac{\partial F}{\partial V}\right)_T = 0.$$ (6.20)

From here, one can obtain the following formula for the relative change of the body volume:

$$\frac{\Delta V}{V} = -\varkappa k_B T \sum_{qj} \frac{1}{\omega_j(\mathbf{q})} \left[\frac{\partial \omega_j(\mathbf{q})}{\partial V}\right].$$ (6.21)

Then, in the framework of the considered quasiharmonic approximation, one of the important parameters of anharmonicity — the Gruneisen coefficient γ_G — is introduced.

Usually, it is assumed that the relative change of the frequency of atomic vibrations as the temperature increases is one and the same for each oscillatory mode and is proportional to the relative change of the volume. Hence,

$$\frac{\Delta\omega}{\omega} = -\gamma_G \frac{\Delta V}{V} = -\gamma_G \varkappa T.$$ (6.22)

The Gruneisen coefficient represents the rate of variation of the natural logarithm of the frequency caused by the variation of the logarithm of the volume. Strictly speaking, it depends on the oscillatory mode:

$$r_j(\mathbf{q}) = -\frac{d\ln\omega_j(\mathbf{q})}{d\ln V}.$$ (6.23)

According to the same theory, the force constants vary, as the temperature rises, in one and the same way:

$$\frac{\alpha_k}{\alpha_{0k}} = \frac{\beta_k}{\beta_{0k}} = \frac{\gamma_k}{\gamma_{0k}} = 1 - 2\gamma_G \varkappa T,$$ (6.24)

where the constants in the denominators, containing subscripts with zero, are related to the harmonic model.

6.4. Scattering of X-rays by anharmonic crystal

The deviation of the crystal behaviour from the ideal harmonic one can be seen in its properties of different scattering of X-rays and thermal neutrons. The analysis of the connection between the parameters of anharmonicity and the laws of the scattering is of great importance.

Consider the formulae of scattering successively for the three approximations [85].

According to the kinematic theory of diffraction, the scattering amplitude can be written as

$$Y(\mathbf{S}) = \sum_{l,k=1}^{N,n} f_k e^{i\mathbf{S}[\mathbf{r}(lk)+\mathbf{u}(lk)]}, \qquad (6.25)$$

where $\mathbf{S}$ is the scattering vector. Its modulus is equal to $(4\pi \sin\theta)/\lambda$. The sum in equation (6.25) is taken over all n atoms in the elementary cell $(1 \leq k \leq n)$ and over all N cells in the crystal $(1 \leq l \leq N)$; f_k is the factor of atomic scattering of the kth atom, which is at rest. The vector $\mathbf{u}(lk)$ denotes the displacement of the atom l, k from the equilibrium position.

The amplitude of scattering can be expressed as YY^*:

$$Y^2 = \sum_{lk}\sum_{l'k'} f_k f_{k'} e^{i\mathbf{S}[\mathbf{r}(lk)-\mathbf{r}'(l'k')]} \left\langle e^{i\mathbf{S}[\mathbf{u}(lk)-\mathbf{u}'(l'k')]} \right\rangle. \qquad (6.26)$$

The displacements of atoms depend on the time; the angular brackets show the averaging over a time period, which is large in comparison with the period of atomic vibrations.

Different processes of scattering of X-rays can be divided into elastic scattering (of zero order) and nonelastic scattering, which goes on with the change of the energy of quanta of normal vibrations (phonons) in the crystal.

For the cubic crystal, the displacements $\overline{u^2}$ are isotropic and the temperature factor for the kth atom is represented by the following formulae:

$$W_k = 8\pi^2 \overline{u^2}(lk) \frac{\sin^2 \theta}{\lambda^2}, \qquad (6.27)$$

$$F(\mathbf{Q}) = \sum_{k=1}^{n} f_k e^{-W_k(\mathbf{Q})} e^{i\mathbf{Q}\mathbf{r}(lk)}, \qquad (6.28)$$

$$Y^2 = N v_z \sum |F(\mathbf{B})|^2 \delta(\mathbf{Q} - \mathbf{B}). \qquad (6.29)$$

Here, the δ-function represents the condition of reflection. It is equal to one if $\mathbf{Q} = \mathbf{B}$ and zero otherwise. The structural factor is represented by $F(\mathbf{Q})$. The thermal factor of intensity of the scattered radiation of Debye–Waller has the form $\exp(-2W_k)$.

The contribution of every oscillatory mode of cubic crystal into the mean-square atomic displacement in any direction is directly proportional to the quantity

$$\frac{E_j(q)}{\omega_j^2(q)},$$

where $E_j(q)$ is the energy of the oscillatory mode j. For high temperatures, the energies are uniformly distributed between the oscillatory modes and $E_j(q) = k_B T$ and $W_k \sim T$. Thus, in the harmonic approximation, W_k contains a term proportional to the temperature.

It turns out that, in the quasiharmonic approach, the quantity W contains linear as well as quadratic terms with respect to the temperature. The anharmonic approach takes into account the cubic term with respect to the temperature as well. For comparison, we write all three these formulae. Recall that the Debye–Waller factor is equal to

$$\exp\left[-2B_k \frac{\sin^2 \theta}{\lambda^2}\right].$$

The harmonic approximation:

$$B_{k\ \text{harm}} = \frac{8\pi^2 k_B T}{\alpha_{0k}} = 8\pi^2 \overline{u^2}(lk). \qquad (6.30)$$

The quasiharmonic approximation:

$$B_{k\ \text{qharm}} = \frac{(8\pi^2 k_B T)(1 + 2\gamma_G \varkappa T)}{\alpha_{0k}}. \qquad (6.31)$$

The anharmonic approximation:

$$B_{k\ \text{anharm}} = \frac{(8\pi^2 k_B T)(1 + 2\gamma_G \varkappa T - 20 k_B T \gamma_{0k}/\alpha_{0k}^2)}{\alpha_{0k}}. \qquad (6.32)$$

TABLE 1. The constants α (J/m^2) and β (10^{10} J/m^3) for some materials

Material	Constant	Value	Source
Si	α	78.5	
	β	−80.9	[38]
Ge	α	57.9	
	β	−59.6	
	α_{Cu}	7.4	
	α_{Cl}	13.5	[90]
	β_{Cu}	−11.5	
CuCl	β_{Cl}	0.0	
	α_{Cu}	6.7	
	α_{Cl}	13.5	[91]
	β_{Cu}	−7.9	
	β_{Cl}	−1.8	
	α_{Ba}	55.4	
BaF$_2$	α_F	35.9	[85]
	β_F	−34.8	

Thus, the intensity of the interferences due to the anharmonicity of atomic vibrations weakens much more than for harmonic crystals.

6.5. Experimental data

Dawson and Willis [38] studied silicon and germanium in the approximation of one-particle anharmonic potential:

$$\Phi_j(r) = \Phi_0 + \frac{1}{2}\alpha_j\left(u_1^2 + u_2^2 + u_3^2\right) + \beta_j u_1 u_2 u_3. \tag{6.33}$$

In Table 1, the results of measurements of the constants α and β for these two semiconductors are presented. It turns out that the anharmonic effects are quite measurable.

Bednarz and Field studied the anharmonic effects in potassium by the X-ray method [86]. The authors emphasize that this metal is an convenient object for investigation. Potassium is a soft metal with a low melting point. Crystallizing in a body-centred crystal lattice, it is characterized by considerable anisotropy. The mean-square displacements of atoms at the room temperature is quite large:

74 pm. From the viewpoint of the electron structure, potassium is an ideal metal with completely delocalized valent electrons and with the spherical Fermi surface.

As has been accepted, it is expedient to represent the expansion of the potential energy in the following form:

$$\Phi(u) = \frac{1}{2}\alpha u^2 + \gamma u^4 + \delta \left(u_1^4 + u_2^4 + u_3^4 - \frac{3}{5}u^4 \right). \tag{6.34}$$

Here, we use the notation $u^2 = u_1^2 + u_2^2 + u_3^2$. While analyzing the data, four models were studied, which differ by the value of the Gruneisen constant.

It turns out that the constant of equation (6.33) are the following:

$$\alpha = 2.42 \text{ J/m}^2; \quad \gamma = -7.20 \cdot 10^{18} \text{ J/m}^4; \quad \delta = 6.56 \cdot 10^{18} \text{ J/m}^4.$$

In general, the results show a noticeable contribution of the parameter δ that characterizes the degree of anisotropy of vibrations in potassium. The contribution of the parameter γ is also essential. The fact that the parameter of anharmonicity δ is positive means that the amplitude of the atomic vibrations in the direction of the nearest neighbours [111] is larger than that in the direction of the second neighbours [100]. In this paper, the conjecture is made that the observed anisotropy of the vibrations is caused by the rather high concentration of vacancies in the crystal lattice of potassium. The latter is typical for the metal under consideration. For instance, near the melting point, the concentration of vacancies reaches 10^{-3}. The vacancy formation is accompanied by the relaxation of neighbouring atoms in the vicinity of the nonoccupied node. The relaxation can reach 5–7% of the interatomic distance. The activation energy of the vacancy diffusion is rather small ($1.02 \cdot 10^{-20}$ J); therefore, the vacancies in potassium are very mobile.

At the melting point, the mean-square amplitude of atomic vibrations (determined through the parameter α) is 17.5% of the interatomic distance.

In paper [87], the experimental data of the X-ray diffraction study of cadmium crystals are analyzed. It is known that cadmium crystallizes as the close-packed hexagonal structure with the axes ratio $c/a = 1.89$ and two atoms in the elementary cell. Four models were studied, among them harmonic and anharmonic. The analysis allows one to conclude that the anharmonic models correlate with the experimental data much better. The relation between the found values of

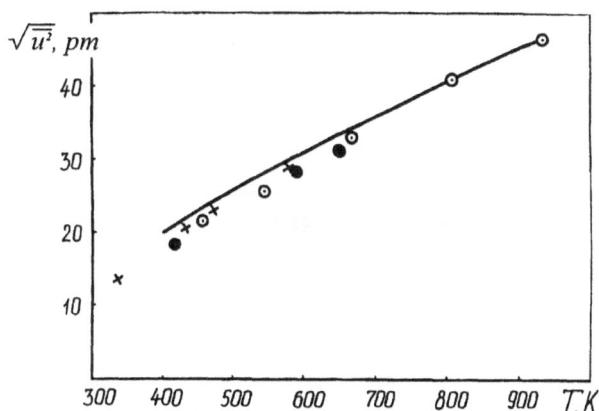

FIGURE 6.2. The influence of the temperature on the mean-square amplitude of atomic vibrations in potassium chloride.
o — reflex (400); • — (600); × — (800). The solid curve is the result of calculation according to the anharmonic law.

the constants shows that the vibrations in the plane of the crystal lattice basis are asymmetric. In cadmium, the potential barrier is lower and the vibrations amplitude is larger in the direction of octahedral internodes of the elementary cell; in the direction of tetrahedral internodes, the barrier is higher and the amplitude is accordingly less. The anisotropy of atomic vibrations along the axes is also fairly large. Along the c-axis, it is equal to 20.2 pm and is equal only to 11.7 pm in the basis plane, along the a-axis.

In Fig. 6.2, experimental points of the temperature dependence of the mean-square atomic displacements for potassium chloride are presented (the data of measurements of integral intensities by James and Brindley, cited from [85]). The atomic masses of ions of K and Cl are rather close to each other; the scattering factors are also close. The solid curve is calculated according to the anharmonic theory

$$\overline{u^2}(T) = \overline{u^2}(T)_{\text{harm}} \left[1 + T\left(2\gamma_G \varkappa - 20k_B\gamma_0/\alpha_0^2\right)\right]. \qquad (6.35)$$

The value of the ratio γ_0/α_0^2 turns out to be equal to $0.14 \cdot 10^{19}$ J^{-1}. Estimations show that the contribution of the third term on the right-hand side of the equation is noticeable if the temperature exceeds 700 K. Accordingly, the best agreement of the anharmonic theory

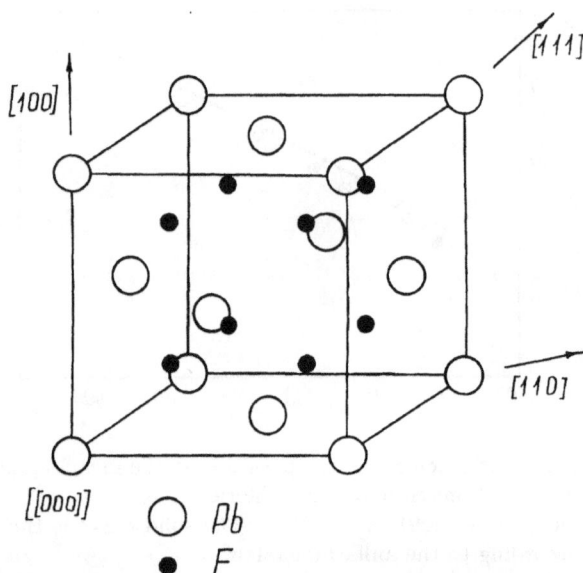

FIGURE 6.3. The elementary cell of the compound β-PbF$_2$.

with the experimental data is observed in the temperature range 700–1000 K. For relatively low temperatures, this theory gives overstated results. At the same time, in the interval 300–700 K, the linear relation between the two quantities described by the formula

$$\overline{u^2} = \frac{3k_B T}{\alpha_0}$$

is clearly observed. The value α_0 turns out to be equal to 25.4 J/m^2.

The study of lead fluoride β-PbF$_2$ is described [88]. This composition shows the electric conductivity thanks to fluorine anions. The ionic conductivity is closely connected with the anharmonicity of atomic vibrations. Lead fluoride has the cubic crystal lattice of the fluorite type (Fig. 6.3). As a result of the systematic study of the monocrystal by the method of X-ray diffraction measurement of the interference intensities, the authors constructed the probabilistic functions of the density distribution for ions of lead and fluorine for different temperatures. The results are represented as maps–cross-sections, on which the equiprobabilistic lines are drawn at certain

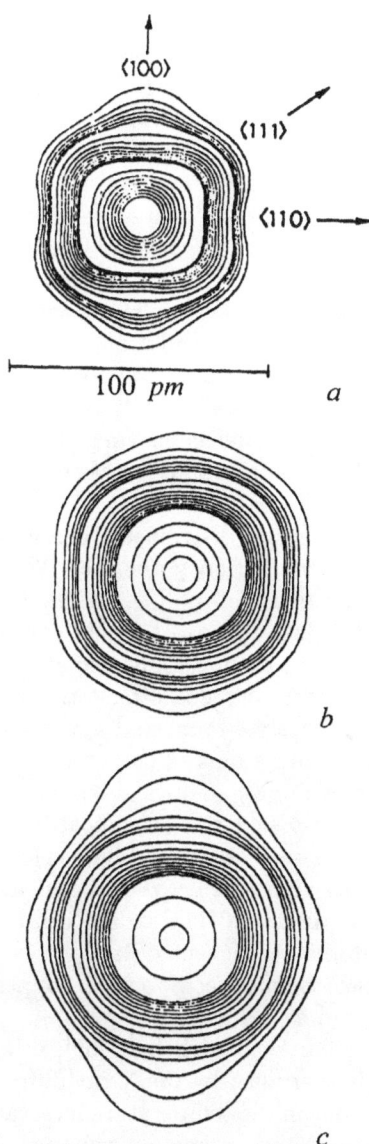

FIGURE 6.4. The cross-sections of the probability functions for the lead ion located at the origin. The experimental data of the authors [88]: The lines of equal density are drawn from 1 to 10 nm^{-1} with the step 2 nm^{-1}; from 20 to 100 nm^{-1} with the step 10 nm^{-1}; from 200 with the step 100 nm^{-1}. The temperatures: a — 295; b — 461; c — 625 K.

distances. From these data, it is convenient to evaluate the anisotropy of atomic vibrations in lead fluoride. Figure 6.4 shows the maps of the probability of positions of lead ions Pb^{2+} in the plane (110). They imply that the largest amplitude of vibrations of this ion is observed along [100]; in other words, in the direction of the lead ion which is the second neighbour. At room temperature, the least amplitude is in the direction [110], where the nearest lead ion is located. As the temperature rises from room temperature to 625 K, the thermal displacements increase; however, these displacements are not the same for different directions. In these conditions, the lowest amplitudes of displacements of the considered lead atom are in the direction of the interatomic bonds Pb^{2+}–F^-, along the solid diagonal [111]. Thus, one can see that the interatomic bonds between atoms of different kinds are the strongest. At the same time, the vibration amplitudes are largest in the direction of bonds between the second neighbours Pb^{2+}–Pb^{2+} [100].

The maps of the density distribution enable us to calculate the so-called one-particle interaction potential. We have seen a similar schematic dependence in Fig. 6.1. The curves $\Phi(r)$ obtained on the basis of experimental data for lead fluoride can be seen in Fig. 6.5. The anisotropy of the interatomic bonds and atomic vibrations can be clearly seen. For neighbouring lead atoms, the curve begins from an amplitude of approximately 40 pm, and appears more gently sloped than for the neighbouring atoms of lead–fluorine. The situation for the fluorine atom shown in Fig. 6.5b is also obvious. Recall that when the curve appears gently sloped, this indicates a weaker bond. The strongest bonds are between atoms of different kinds. It is interesting that the fluorine ions vibrate with the largest amplitude along the solid diagonal in the direction [111] to the centre of the elementary cell; there are no atoms at the centre itself.

The data for two chemical compounds, copper chloride and barium fluoride, are given in Table 1.

In the paper by Zucker and Schulz [89], a fast ionic conductor lithium nitride Li_3N is studied by the X-ray diffraction method. The results of investigations in the temperature interval from 294 to 888 K enable the authors to detect strong anharmonic effects.

Lithium nitride crystallizes as hexagonal syngony with the axes $a = 36.5$ and $b = 38.8$ nm. The nitrogen atom occupies the centre of the elementary cell and is surrounded by eight lithium atoms; the

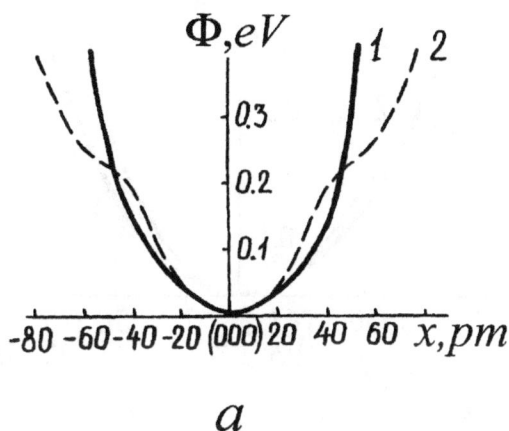

a

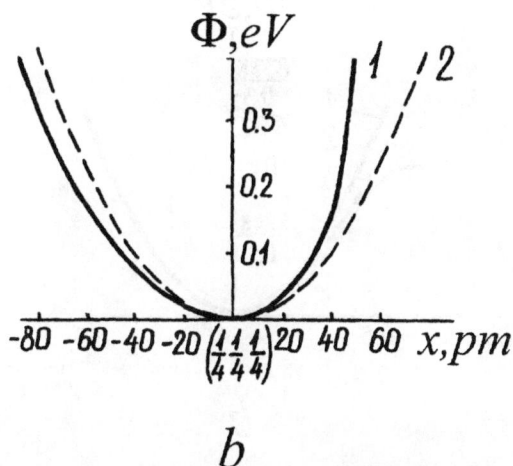

b

FIGURE 6.5. The curves of the potential energy of lead fluoride in the one-particle approximation:

a — the dependence of the potential energy of interaction on the distance for the atom of lead at the origin [[000]]; 1 — the energy of the bond with the nearest fluorine ion in the direction [111]; 2 — the energy of the bond with the lead ion (the second neighbour) in the direction [100].

b — the dependence $\Phi(r)$ for the fluorine ion [[$\frac{1}{4}\frac{1}{4}\frac{1}{4}$]]; 1 — the energy of the bond with the lead ion along [111]; 2 — the energy of the bond with the fluorine ion in the direction [100].

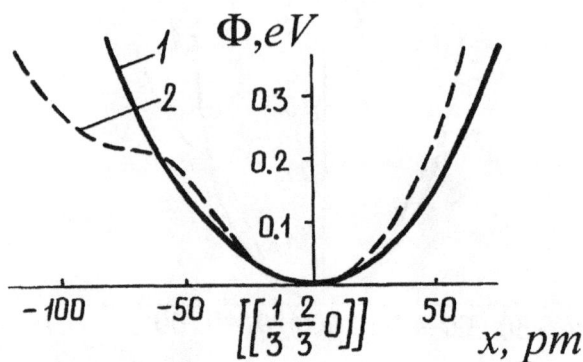

a

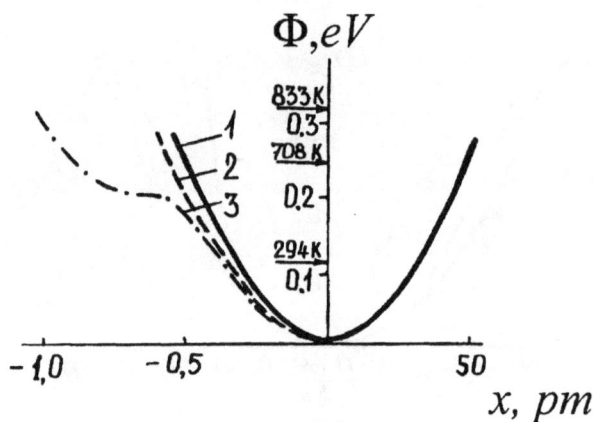

b

FIGURE 6.6. The curves of the potential energy of the interatomic interaction in ionic compound Li$_3$N (data of the authors of [89]):
a — the vibrations of the ion Li(2) in the direction of the atomic chain
N–Li(2)–Li(2)–N; 1 — the harmonic model; 2 — the anharmonic model,
the temperature 888 K. *b*— the influence of the temperature onto the form
of dependence $\Phi(r)$; 1 — 294; 2 — 708; 3 — 833; 4 — 888 K.

compound forms ionic bonds. It is assumed that high ionic electric conductivity is caused by the jumps of ions Li(2). The displacements of ions are favoured by the considerable anharmonicity of atomic vibrations. This can be seen on the curve $\Phi(r)$ for ion Li(2) (Fig, 6.6b). The attention should be paid to the asymmetry of the potential curve: it goes slopingly when the neighbouring atoms are moving apart, the variation of the interaction energy is weaker than when the atoms approach each other.

19 parameters are needed to expand the potential energy into the series; the experimental data is in good agreement with the calculated results.

The mean-square displacements of ions Li(2) show a noticeable anisotropy: along the c-axis they are essentially larger than in the perpendicular direction (Fig. 6.6).

The expansion of the potential energy for cuprous chloride CuCl was used in the form similar to (6.32). This compound was studied by the neutron diffraction method [90] as well as by the X-ray diffraction one [91]. The parameters of harmonicity and anharmonicity are given in Table 1 mentioned above. Data of the anharmonic model are in good agreement with the experiment.

The semiconductive compound CdTe has the structure of zinc blende: every atom of cadmium is surrounded by four atoms of tellurium and vice versa. It has been established [52] that Cd-atoms vibrate with larger amplitudes than Te-atoms and to greater degree are exposed to the anharmonic effects. The average values of the parameters measured by the authors are $\alpha_{Cd} = 20.8$; $\alpha_{Te} = 24.5$ J/m^2; $\gamma_{Cd} = 3.36 \cdot 10^{20}$ J/m^4.

The shift of frequencies of normal vibrations in the crystal lattice of niobium is observed as the temperature rises in the interval from 293 to 2223 K [46]. This indicates the anharmonicity property of this refractory metal.

The frequencies of the transverse mode $[\zeta\zeta 0]$ also diminish in the alloys Fe$_{75}$Si$_{25}$ and Fe$_{80}$Si$_{20}$ as the temperature rises from 293 to 1203 K [93].

CONCLUSION

We have considered the characteristics of coordinated oscillatory motion of atoms in solids. Many properties of materials are stipulated by the crystal lattice dynamics.

The basic tool for the measurement of the mean-square amplitude of vibrations is X-ray diffraction. Correct data about the frequency spectrum is provided with the help of the technique of neutron scattering by solids. In general, the oscillatory behavior of a real three-dimensional crystal lattice is similar to the behavior of a model one-dimensional lattice. The higher is the frequency of the ith oscillatory mode, the less is its mean-square amplitude. The magnitude of the atom vibration amplitude and its temperature dependence are a measure of the energy of interatomic bonds.

Dissolution of atoms of additions in metals, intermetallides and carbides results in substantial change of the mean-square amplitude. Obviously, this is caused by the influence of these atoms on the bond energy and the frequency spectrum.

In semiconductors and semiconductive compounds, the increase of the amplitude of atoms vibrations is connected with the delocalization of electrons of the valence shells and with the decrease of the width of the forbidden band. The amplitudes increase as the degree of the ionicity of interatomic bonds in chemical compounds increases.

The lattice oscillatory mode, which propagates in a symmetric crystallographic direction, corresponds to oscillations of atomic planes, as a whole, in the direction parallel to the wave vector. This idea due to Foreman and Lomer turns out to be very fruitful. On its basis, with the help of the Fourier analysis, one can calculate the interplane force constants as well as the velocities of the elastic waves of the lattice and their dependence on the wave vector.

Theory of Born and Karman is widely used for analysis of interactions between atoms in the vibration processes. According to this theory, an atom under consideration interacts with every atom in pairs; the restoring force is directly proportional to the relative displacement of atoms. The Born and Karman analysis seems to be simple, logical and obvious. However, in most cases, it does not ensure a satisfactory fit with experimental data. Authors of numerous models of interatomic bonds are compelled to introduce some additional parameters. Unfortunately, most of these parameters have no definite physical sense.

In modern physics of solids, a great attention is paid to constructing and studying the frequency spectra of solids.

Many investigations are devoted to the anharmonicity of atomic vibrations and to their anisotropy. Anharmonicity results in decrease of the frequencies ("softening" of the oscillatory modes) and in increase of the mean-square amplitudes.

The author believes that further experimental investigations of crystal lattices dynamics will allow us to achieve the best understanding of the physical nature of atomic interactions in various real solids.

ACKNOWLEDGEMENTS

My sincere gratitude to my wife Lydia for her support during the work on the book.

I would like to express my deep gratitude to my co-workers L.K. Orzhitskaya and L.I. Pospelova for many years of participation in experiments.

I am also grateful to E.V. Pankratiev for preparation of the English version of the text and to Yu.S. Pavlyutkin and S.A. Khaduta who assisted me with the illustrations.

BIBLIOGRAPHY

1. Maradudin A.A., Montroll E.W., Weiss G.H., and Ipatora I.P. *Theory of Lattice Dynamics in the Harmonic Approximation*, Academic Press, N.-Y., 1971.
2. Reissland J.A. *The Physics of Phonons*, John Wiley and sons Ltd., London–New-York–Sydney–Toronto, 1973.
3. Brillouin L. et Parodi M. (editeurs) *Propagation des Ondes dans les Milieux Périodiques*, Masson et C., Paris, 1956.
4. Cochran W. *The Dynamics of Atoms in Crystals*, Arnold Ld., London, 1973.
5. Brüesch P. *Phonons: Theory and Experiments. I. Lattice Dynamics and Models of Interatomic Forces*, Springer-Verlag, Berlin–Heidelberg–New York, 1982.
6. Born M. and Huang K. *Dynamical Theory of Crystal Lattices*, Oxford, 1985.
7. Stassis C. Lattice dynamics. *Methods of Experimental Physics*, Academic Press, 1986, Vol. 23, Part A, 369–440.
8. James R.W. *The Optical Principles of the Diffraction of X-Rays*, London, 1950.
9. Il'ina V.I., Kritskaya V.K., and Kurdyumov G.V. *Investigation of the value and character of distortions of the crystal lattice of martensite*, Probl. Met-alloved. Fiz. Met. The third collection of proceedings of IMF TsNIIChM, 1952, Metallurgizdat, Moscow, 1952, 95–99.
10. Kurdyumov G.V., Il'ina V.I., Kritskaya V.K., and Lysak L.I. *X-ray diffraction study of distortions and of the bond forces in crystal lattices of metals and alloys*, Probl. Metalloved. Fiz. Met. The fourth collection of proceedings of IMF TsNIIChM, 1955, Metallurgizdat, Moscow, 1955, 339–359.
11. Chipman D.R. and Paskin A. *Temperature diffuse scattering of X-rays in cubic powders*, J. Appl. Physics, **30**, no. 12, (1959), 1998–2001.
12. Guinier A. *Théorie et Technique de la Radiocristallographie*, Paris, 1956.
13. Orzhitskaya L.K., Levitin V.V., Pospelova L.I., and Kurnavina L.P. *Influence of alloying elements onto the thermal vibrations of atoms in carbide phase on the base of VC*, Fiz. Met. Metalloved., **54**, no. 1, (1982), 101–109.
14. Brockhouse B.N., Arase T., Caglioti G., Rao K.R., Woods A.D.B. *Crystal dynamics of lead. I. Dispersion curves at 100 K*, Physical Review, **128**, no. 3, (1962), 1099–1111.

15. Woods A.D.B., Brockhouse B.N., March R.H., and Stewart A.T. *Crystal dynamics of sodium at 90 K*, Physical Review, **128**, no. 3, (1962), 1112–1120.

16. Il'ina V.I., Kritskaya V.K., Kurdyumov G.V., Osip'yan Yu.A., and Stelletskaya T.I. *Investigation of the dependence of the bond forces on the state of crystals of metals and solid solutions*, Probl. Metalloved. Fiz. Met. IMF TsNIIChM, 1958, Metallurgizdat, Moscow, 1958, 462–484.

17. Zhurkov S.N., Betekhtin V.I., and Slutsker A.I. *Desorientation of blocks and strength of metals*, Fiz. Tverd. Tela, **5**, no. 5, (1963), 1326–1333.

18. Gupta O.P. *Temperature dependence of the anharmonic Debye–Waller factor for cubic metals*, J. Phys. Soc. Japan, **52**, no. 12, (1983), 4237–4247.

19. Reddy S.V. and Suryanarayana S.V. *X-ray determination and Debye temperature of some ternary silver-base alloys*, Indian J. Pure Appl. Physics, **22**, no. 3, (1984), 161–163.

20. Bhikshamaiah G. and Suryanarayana S.V. *Lattice parameters and Debye temperatures of Ag–Cd–Zn (α-phase) alloys*, J. Less-Common Metals, **132**, no. 1, (1987), 29–35.

21. Dutchak Ya.I. and Chekh V.G. *High-temperature X-ray study of dynamics of the lattice of compounds AlCo and AlNi*, Zh. Fiz. Khim., **55**, no. 9, (1981), 2342–2345.

22. Pathak P.D. and Desai R.J. *Thermal properties of some H.C.P. metals*, Physica status solidi, **64**, no. 2, (1981), 741–745.

23. Gopikrishna N., Sirdeshmukh D.R., and Gschneidner K.A. *X-ray determination of mean-square amplitudes of vibration and associated Debye temperatures of scandium and terbium*, Indian J. Pure Appl. Physics, **26**, no. 12, (1988), 724–725.

24. Sirota N.N., Zhabko T.E., and Orlova N.S. *On anisotropy of thermal vibrations of atoms of titanium*, Dokl. Akad. Nauk BSSR, **30**, no. 9, (1986), 793–795.

25. Gopikrishna N., Sirdeshmukh D.B., Ramarao B., Beaudry B.J., Gschneidner K.A., Jr. *Mean square amplitudes of vibration and associated Debye temperatures of dysprosium, gadolinium, lutenium and yttrium*, Indian J. Pure Appl. Physics, **24**, no. 7, (1986), 324–326.

26. Gopikrishna N. and Sirdeshmukh D.B. *Mean-square amplitudes of vibration and associated Debye temperatures of erbium*, Physica status solidi, **B116**, no. 2, (1983), K105–K108.

27. Prytkin V.V. and Biblik E.V. *Mean-square displacements of atoms along the principal directions of cadmium monocrystal in the interval 78–300 K*, Metallofiz., **13**, no. 3, (1991), 14–20.

28. Levitin V.V. and Orzhitskaya L.K. *On the mechanism of creep of high-temperature alloys*, Izv. Akad. Nauk SSSR, Metally, no. 5, (1978), 94–102.

29. Tewary V.K. *On a relation between the monovacancy formation energy and the Debye temperature for metals*, J. Physics F, **3**, no. 5, (1973), 704–708.

30. Bokstein B.S., Bokstein S.Z., and Zhukhovitskij A.A. *Thermodynamics and kinetics of diffusion in solids*, Metallurgiya, Moscow, 1974.

31. Pauling L. *The nature of chemical bonds*, Goskhimizdat, Moscow – Leningrad, 1947.

32. Slater J.C. *Atomic shielding constants*, Physical Review, **36**, no. 1, (1930), 57–64.
33. Kowlsohn Ch. *Valency*, Mir, Moscow, 1982
34. Kittel Ch. *Introduction to Solid State Physics*, John Wiley & sons, Inc, Chapman & Hall, Ltd., London, 1956.
35. Ashcroft N.W. and Mermin N.D. *Solid State Physics*, Holt, Rinehart and Winston, New-York–Chicago–San Francisco, 1976.
36. Batterman B.W. and Chipman D.R. *Vibrational amplitudes in germanium and silicon*, Physical Review, **127**, no. 3, (1962), 690–693.
37. Bilderback D.H. and Colella R. *Valence charge density in grey tin: X-ray determination of the (222) "forbidden" reflection and its temperature dependence*, Physical Review B, **11**, no. 4, (1975), 793–797.
38. Dawson B. and Willis B.T.M. *Anharmonic vibration and forbidden reflexions in silicon and germanium*, Proceeding of the Royal Society. **298**, no. 1454, (1967), 307–315.
39. Roberto J.B., Batterman B.W., and Keating D.T. *Diffraction studies of the (222) reflection in Ge and Si: anharmonicity and the bonding electrons*, Physical Review B, **9**, no. 6, (1974), 2590–2599.
40. Fujimoto I. *Temperature and pressure dependence of the Si (222) forbidden reflection and the vibration of the bonding charge*, Physical Review B, **9**, (1974), 591–599.
41. Tulinov A.F. *Influence of crystal lattice on some atomic and nuclear processes*, Usp. Fiz. Nauk, **87**, no. 4, (1985), 585–598.
42. Bazylev V.A. and Zhevago N.K. Generation of intensive electromagnetic radiation by relativistic particles, Usp. Fiz. Nauk, **137**, no. 4, (1982), 605–662.
43. Appleton B.R. , Erginsoy C., and Gibson W.M. *Channeling effects in the energy loss of 3–11 MeV protons in silicon and germanium single crystals*, Physical Review, **161**, no. 2, (1967), 330–349.
44. Phillips J.C. *Spectra and specific heats of diamond-type lattices*, Physical Review, **113**, no. 1, (1959), 147–155.
45. Bilderback D.H. and Colella R. *X-ray determination of valence-electron charge density and its temperature dependence in indium antimonide*, Physical Review B. Solid State, **13**, n. 6, (1976), 2479–2487.
46. Guthoff F., Henion B., Herzig C. a.o. *Lattice dynamics and selfdiffusion in niobium at elevated temperatures*, J. Physics: Condenced Materials, **6**, no. 31, (1994), 6211–6220.
47. Kyutt R.N. *Mean-square displacements of atoms and the Debye temperatures of crystals $A^{III}B^V$*, Fiz. Tverd. Tela, **20**, no. 2, (1978), 395–398.
48. Sirota N.N. and Sidorov A.A. *Temperature dependence of the intensity of diffraction maxima on X-ray pictures of semiconductive compounds GaAs, InAs, InP in the temperature range 7–310 K*, Dokl. AN SSSR, **280**, no. 2, (1985), 352–356.
49. Colella R. *Experimental investigations of bonding charges in diamond and zinc blende structures*, Physica Scripta, **15**, no. 2, (1977), 143–146.
50. Colella R. *X-ray investigation of bond-charge density in gallium arsenide*, Physical Review B, **3**, no. 12, (1971), 4308–4311.

51. Attard A.E., Mifsud F.A., Sant A.K., and Sultana J.A. *Covalent charge transfer in III-Y compounds*, J. Physics C, **2**, no. 5, (1969), 816–823.

52. Horning R.D. and Staudenman J.L. *X-ray vibrational studies on (100)-oriented CdTe crystals as a function of the temperature (8–350 K)*, Physical Review B. Condensed Materials, **34**, no. 6, (1986), 3970–3979.

53. Bashir J., Butt N.M., and Khan Q.H. *Debye–Waller factors and Debye temperature of ZnTe*, Acta Crystallografica, **44**, no. 5, (1988), 638–639.

54. Yakimavichius J.A., Purlus R.P., and Shirvaitis A.J. *Temperature dependence of mean square dynamic displacements and characteristic temperatures of $A^{II}B^{VI}$ compounds*, Physica status solidi, **B110**, no. 1, (1982), K51–K56.

55. Miyake S. and Hoshino S. *Temperature characteristic of X-ray intensity reflection from crystals having zincblende- and wurtzite-type structures*, Reviews of modern physics, **30**, (1958), 172–174.

56. Sakata M. and Hoshino S. *Neutron diffraction study of asymmetric anharmonic vibration of the cuprous chloride*, Acta crystallografica, **A30**, part 5, (1974), 655–661.

57. Valvoda V. and Ječny J. *X-ray diffraction study of anharmonic thermal vibrations in CuCl*, Physica status solidi, **A45**, no. 1, (1978), 269–275.

58. Subhadra K.G. and Sirdeshmukh D.B. *X-ray determination of the mean Debye–Waller factors, amplitudes of vibration and Debye temperature of six crystals with CsCl structure*, Pramana J. Physics, **10**, no. 6, (1978), 597–600.

59. Srinivas K. and Sirdeshmukh D.B. *X-ray determination of the mean amplitudes vibration and Debye temperature of KRS-6 ($Tl_{0.7}Br_{0.3}$)*, Crystal Research and Technology, **21**, no. 9, (1986), K165–166.

60. Srinivas K. and Sirdeshmukh D.B. *X-ray determination of the Debye–Waller factors and Debye temperatures of AgCl and AgBr*, Pramana J. Physics, **23**, no. 5, (1984), 595–597.

61. Bokij G.V. and Poraj-Koshits M.A. *Practical course of X-ray diffraction analysis*, Mosk. Gos. Univ., Moscow, 1951.

62. Goldschmidt H.J. *Interstitial Alloys*, Butterworths. London, 1967.

63. Yarnell J.L., Warren J.L., and Koenig S.H. *Experimental dispersion curves for phonons in aluminum*, Lattice Dynamics. Proceeding of the International Conference. Denmark, 1963. Ed. Wallis P.F. Pergamon Press, Ox.-Ld.-Ed.-N.-Y., 1965, 57–61.

64. Nakagawa Y. and Woods A.D.B. *Lattice dynamics of niobium*, Lattice Dynamics. Proceeding of the International Conference. Denmark, 1963. Ed. Wallis P.F. Pergamon Press, Ox.-Ld.-Ed.-N.-Y., 1965, 39–48.

65. Woods A.D.B. *Lattice dynamics of tantalum*, Physical Review, **136**, no. 3A, (1964), 781–783.

66. Foreman A.J.E. and Lomer W.M. *Lattice vibrations and harmonic forces in solids*, Proceedings of the Physical Society, Sec. B, **70**, part 2, (1957), 1143–1150.

67. Chen S.H. and Brockhouse B.N. *Lattice vibration of tungsten*, Solid State Communications, **2**, no. 3, (1964), 73–77.

68. Higuera B.J. and Brotzen F.R. *Phonon dispersion in Ta-W system*, Physical Review B, **31**, no. 2, (1985), 730–734.

69. Woods A.D.B. and Chen S.H. *Lattice dynamics of molybdenum*, Solid State Communications, **2**, no. 7, (1964), 233–237.
70. Dolling G. and Cowley R.A. *The termodynamic and optical properties of germanium, silicon, diamond and gallium arsenide*, Proceedings of the Physical Society, **88**, no. 560, (1966), 463–494.
71. Phillips J.C. *Vibration spectra and specific heats of diamond-type lattice*, Physical Review, **113**, no. 1, (1959), 147–155.
72. Zarestky J. and Stassis C. *Lattice dynamics of γ-Fe*, Physical Review B, **35**, no. 9, (1987), 4500–4502.
73. Fultz B., Anthony L., Nagel L.J., and Nicklow R.M. a.o. *Phonon densities of states and vibrational entropies of ordered and disordered Ni_3Al*, Physical Review B, **57**, no. 5, (1995), 3315–3321.
74. Bardhan P., Chen H., and Cohen J.B. *Premonitory effects in Cu_3Au near the order-disorder transformation*, Philosophical Magazine, **35**, no. 6, (1977), 1653–1666.
75. Lander G.H. and Brown P.H. *A neutron diffraction study near the order-disorder temperature*, J. Physics C: Solid state physics, **18**, no. 10, (1985), 2017–2024.
76. Robertson I.M. *Phonon dispersion iron–aluminum alloys*, J. Physics: Condensed Materials, **3**, no. 42, (1991), 8181–8194.
77. Ivanov A.S., Katsnelson M.I., and Mikhin A.G. a o. *Phonon spectra, interatomic interaction potentials and simulation of lattice defects in iridium and rhodium*, Philosophical Magazine B, **69**, no. 6, (1994), 1183–1195.
78. Pleschiutschnig J., Blaschko O., and Reichardt W. *Phonon dispersion and heat capacity of scandium*, Physical Review B, **44**, no. 13, (1991), 6795–6798.
79. Pleschiutschnig J., Blaschko O., and Reichardt W. *Lattice dynamics of lutetium*, Physical Review B, **41**, no. 2, (1990), 975–979.
80. Powell B.M., Martell P., and Woods A.D.B. *Lattice dynamics of niobium-molybdenum alloys*, Physical Review, **171**, no. 3, (1968), 727–736.
81. Strauch D. and Dorner B. *Phonon dispersion in GaAs*, J. Physics: Condensed Matter, **2**, (1990), 1457–1474.
82. Brout R. *Sum rule for lattice vibrations in ionic crystals*, Physical Review, **113**, no. 1, (1959), 43–44.
83. Rosenstock H.B. *Sum rule for lattice vibrations: application to forces in diamond structures*, Physical Review, **129**, no. 5, (1963), 1959–1961.
84. Rosenstock H.B. and Blanken G. *Interatomic forces in various solids*, Physical Review, **145**, no. 2, (1965), 546–554.
85. Willis B.T.M. *Lattice variation and the accurate determination of structure factors for the elastic scattering of X-rays and neutrons*, Acta crystallographica, **A25**, Part 2, (1969), 277–300.
86. Bednarz B. and Field D.W. *An X-ray diffraction study of potassium*, Acta crystallographica, **A38**, (1982), 3–10.
87. Field D.W. *On anharmonicity in cadmium*, Acta crystallographica, **A38**, (1982), 10–12.
88. Schulz H., Perenthaler E., and Zucker U.H. *Anharmonic thermal vibrations and atomic potentials in lead fluoride ($\beta-PbF_2$) as a function of temperature*, Acta crystallographica, **A38**, (1982), 729–733.

89. Zucker U.H. and Schulz H. *Statistical approaches for the treatment of anharmonic motion in crystals. 2. Anharmonic thermal vibrations and effective atomic potentials in the fast ionic conductor lithium nitride ($Li_3 N$)*, Acta crystallographica, **A38**, Part 5, (1982), 568–576.

90. Sakata M. and Hoshino S. *Neutron diffraction study of asymmetric anharmonic vibration of the copper atom in cuprous chloride*, Acta crystallographica, **A30**, (1974), 655–661.

91. Valvoda V. and Ječny J. *X-ray diffraction study of anharmonic thermal vibration in CuCl*, Physica status solidi (a), **15**, (1978),269–275.

92. Sears V.F., Svenson E.C., and Powell B.M. *Phonon density of states in vanadium*, Canadian J. Physics **73**, no. 11–12, (1995), 726–734.

93. Randl O.G., Vogl G., and Petry W. a.o. *Lattice dynamics and related diffusion properties of intermetallics. I. $Fe_3 Si$*, J. Physics: Condenced Materials, **7**, no. 30, (1995), 5983–5999.

Index

PHYSICS REVIEWS

Editor:
I.M. Khalatnikov
L.D. Landau Institute
of Theoretical Physics
Russian Academy of Sciences
Moscow

Aims and Scope

Physics Reviews reports on significant developments in physics research and presents reviews by scientists from the former Soviet Union and will be of particular interest to research scientists who do not read Russian.

World Wide Web Addresses

Additional information is also available through the Publisher's web home page site at http://www.cambridgescientificpublishers.com

Ordering Information

Each volume consists of an irregular number of parts depending upon extent. Issues are available individually as well as by subscription. 2004 Volume 21/22.

Orders may be placed with your usual supplier or at the address shown below. Journal subscriptions are sold on a per volume basis only. Claims for nonreceipt of issues will be honored if made within three months of publication of the issue. All issues are dispatched by airmail throughout the world.

Subscription Rates

Base list subscription price per volume: EUR 120.00.* This price is available only to individuals whose library subscribes to the journal OR who warrant that the journal is for their own use and provide a home address for mailing. Orders must be sent directly to the Publisher and payment must be made by check or credit card.

Separate rates apply to academic and corporate/government institutions.

EUR (Euro). The Euro is the worldwide base list currency rate. All other currency payments should be made using the current conversion rate set by Publisher. Subscribers should contact their agents of the Publisher. All prices are subject to change without notice.

Orders should be placed through the Publisher at the following address:

Cambridge Scientific Publishers Ltd
PO Box 806
Cottenham
Cambridge
CB4 8RT
UK
Tel: +44 (0)1954 251283
Fax: +44 (0)1954 252517
Email: janie.wardle@cambridgescientificpublishers.com
Website: www.cambridgescientificpublishers.com

Printed in UK.

August 2004

Physics Reviews

Notes for Contributors

Manuscripts

Papers should be typed with double spacing and wide margins (3 cm) on good quality paper and submitted in triplicate. Authors may also submit papers on disk in any format. Papers should be sent to I.M. Khalatnikov, L.D. Landau Institute of Theoretical Physics, Kosygin Str. 2, Moscow, 117940, Russia or to Janie Wardle, Cambridge Scientific Publishers, P.O. Box 806, Cottenham, Cambridge, CB4 8RT, UK. Email: janiewardle@cambridgescientificpublishers.com.

Submission of a paper to *Physics Reviews* will be taken to imply that it represents original work not previously published, that is not being considered for publication elsewhere, and that if accepted for publication it will not be published in the same language without the consent of the Editors and Publisher.

Language: The language of publication is English.

Abstract: Each paper requires an abstract of 100-150 words summarizing the significant coverage and findings. It is a condition of acceptance by the Editor of a typescript for publication that the Publishers acquire automatically the copyright in the typescript throughout the world.

Figures

All figures should be numbered with consecutive arabic numbers, have descriptive captions and be mentioned in the text. Keep figures separate from the text, but indicate an approximate position for each in the margin.

Preparation: Figures submitted must be of a high enough standard for direct reproduction. Line drawings should be prepared in black (India) ink on white paper or on tracing cloth, with all the lettering and symbols included. Alternatively, good sharp photoprints ("glossies") are acceptable. Photographs intended for half-tone reproduction must be glossy original prints of maximum contrast. Clearly label each figure with author's name and figure number, indicate top" where this is not obvious. Redrawing or retouching of unusable figures will be charged to the authors.

Size: Figures should be planned so that they reduce to 12.7 cm column width. The preferred width of line drawings is 15 to 22 cm with capital lettering 4 mm high, for reduction by one-half. Photographs for half-tone reproduction should be about twice the desired size.

Color plates: Whenever the use of color is an integral part of the research, or where the work is generated in color, the Journal will publish the color illustrations without charge to the author. Reprints in color will carry a surcharge. Please write to the Publisher for details.

Equations and Formulae

Any mathematical or chemical notation should be clearly marked.

Mathematical: Mathematical equations should preferably be typewritten, with subscripts and superscripts clearly shown. It is helpful to identify unusual or ambiguous symbols in the margin when they first occur. To simplify typesetting, please use: 1) the "exp" form of complex exponential functions; 2) fractional exponents instead of root signs; and 3) the solidus (/) to simplify fractions – e.g. exp $x^{1/2}$.

Chemical: Ring formulae and other complex chemical matter are extremely difficult to typeset. Please, therefore, supply reproducible artwork for equations containing such chemistry.

Marking: Where chemistry is straightforward and can be set (e.g. single-line formulae please help the typesetter by distinguishing between, e.g., double bonds and equal signs, an single bonds and hyphens, where there is ambiguity. The printer finds it extremely difficult t identify which symbols should be set in roman (upright) or italic or bold type, especially whe the paper contains both mathematics and chemistry. Therefore, please underline a mathematical symbols to be set in italic and put a wavy line under bold symbols. Other letter not marked will be set in roman type.

Tables
Number tables consecutively with arabic numerals and give a clear descriptive caption at th top. Avoid the use of vertical rules in the tables. Indicate in the margin where the printer shoul place the tables.

References and Notes
References and Notes are indicated in the text by consecutive superior arabic number (without parentheses). The full list should be collected and typed at the end of the paper i numerical order. Listed references should be complete in all details but excluding article title in journals. Authors' initials should follow their names; journal title abbreviations shoul conform to *Physical Abstracts* style. Examples:

Smith, A.B. and Jones, C.D., 1990, *J. Appl. Phys.* 34, 296.

Brown, R.B. *Molecular Spectroscopy*, Gordon and Breach, New York, 1965, 3rd ed., Chap. 6 pp. 95-106.

Proofs
Contributors from the former Soviet Union will receive page proofs (including figures) fo correction via our internal courier network to Moscow. These must be returned to our Moscov office (Victor Selivanov, Lebedev Physical Institute, 53 Leninsky Prospect, Moscow 117924 Russia. Email: victor@vandv.ru) within 48 hours of receipt. All other contributors will receiv page proofs (including figures) by airmail for correction, which must be returned to the printe within 48 hours of receipt. Please ensure that a full postal address if given on the first page o the typescript, so that proofs are not delayed in the post. Author's alterations in excess of 10% of the original composition cost will be charged to authors

Reprints
Additional reprints may be ordered by completing the appropriate form sent with proofs.

Page Charges
There are no page charges to individuals or institutions.